BEYOND MAPPING

BEYOND MAPPING:
CONCEPTS,
ALGORITHMS, AND
ISSUES IN GIS

Joseph K. Berry

1993
GIS World Books

GIS World, Inc.
Fort Collins, Colorado, USA

Publisher: H. Dennison Parker
Vice President, Publishing: Nora Sherwood
Book Projects Manager: Bea Ferrigno
Editorial Assistant: Barbara Bernauer
Production Manager: Christine Thompson
Cover Design: Darin E. Sanders
Layout, Graphics, and
Composition: Darin E. Sanders
Wade L. Smith
Printer: Edward Brothers, Inc.

Library of Congress Cataloging-in-Publication Data

Berry, Joseph K., 1946-
    Beyond mapping : concepts, algorithms, and issues in GIS / Joseph K. Berry.
      266 p. 15x23 cm.
      Collection of columns published 1989-1993.
      Includes bibliographical references and index.
      ISBN 0-9625063-6-2
      1. Geographic information systems. I. Title.
    G70.2.B47 1993
    910' 285—dc20   92-42643
                                                              CIP

ISBN 0-9625063-6-2

ABCDEFGHIJKLMNOP

GIS World Books
400 N. College Ave.
Suite 100
Fort Collins, CO 80524, USA

Contents

Preface

❖   ❖   ❖

My reasons for writing this book are very simple. To many of us, technology seems to be accelerating at hyperspeed. We are hurled down a path at a blinding pace, trying to make sense of the blurred world. By the time you fix on an exit sign, you have already passed it several miles back. Seemingly arcane terminology, dense theory, intricate protocols, and impractical examples fuel this uncomfortable feeling.

Geographic information systems (GIS) technology is more different from than it is similar to traditional mapping. These differences are both an asset and a liability. On one hand, GIS offers powerful new tools to address complex issues in entirely new ways. On the other hand, these new tools are unfamiliar and challenge many of the fundamental concepts we have developed over 8,000 years of mapping.

Geographic information systems technology is struggling in its transition from the researcher to the general user. The user community tends to define GIS in the more comfortable terms of computer mapping and spatial database management. These basic activities are the cornerstones of GIS. They automate our historic map analysis procedures and provide a familiar foothold on the technology. However, this perspective severely limits the full potential of this new technology. It is the understanding of the new procedures that translates into the new approaches to land planning and management needed to address our increasing complex issues.

It is to this understanding that this book is dedicated. It is designed for any student or professional interested in the creative application of GIS. It helps to develop an understanding of the map analysis "toolbox" used in spatial models. A light-hearted style and practical examples are used in conveying the underlying theory.

The book is based on a series of articles in the "Beyond Mapping" column in *GIS WORLD* magazine from 1989 to 1993. Over the years the titles to the articles have been tongue-in-cheek or satirical; therefore, subtitles have been added to reflect the contents of the articles. The articles have been edited and grouped under 10 topics. The topics are ordered to balance what the readers need to develop for a broad understanding of the potentials and pitfalls of GIS, as well as the concepts and procedures of specific map analysis operations. For the most part, these two thrusts are interlaced. Topics 1, 4, 8, and 10 are primarily concerned with the broader issues. Topics 2, 3, 5, 6, 7, and 9 focus mainly on technical aspects. The reader is encouraged to complete the book in the order it is presented. However, there is an internal reference for related discussions in other articles. At the end of each topic, there is a list of recommended readings.

The companion PC-based software, Tutorial Map Analysis Package (tMAP™), provides a series of hands-on exercises corresponding to each of the topics. This practical experience reinforces the map analysis concepts presented. For those readers with the tMAP software,there is a specially designed tutorial corresponding to each topic in this book (TMAP0 through TMAP10 command files [CMD]) in addition to several other related tutorials. Order information and a brief description of tMAP are found in appendix A.

Many people have contributed in many ways to the writing of this book. Thoughts rarely materialize in a vacuum. Discussions with colleagues stimulate the development of concepts and the means to explain them. Notable among these interactions are those which began at Yale University in the mid-1970s with C. Dana Tomlin and Kenneth L. Reed. In a similar sense, books rarely materialize in a vacuum. The continued support and encouragement of H. Dennison Parker provided the means. Neither thoughts nor books can materialize without an environment that embraces the spirit of academics within practical realities of living. Foremost among those providing this basic element are my parents, my wife Joyce, and daughter Ali, whose combined support, patience, and practical insights kept me on track. It is within their love and concern that this book took form.

Note:

The graphics for the figures in this book were created using the tMAP system and its optional software, using a standard personal computer (PC) and screen capture for printing. The low resolution of the tutorial database (25 rows by 25 columns) results in the jagged appearance of the raster images. A higher resolved database would result in higher quality images.

Introduction
An Overview of Basic Terminology and Structure

❖　　　❖　　　❖

Sticks and stones may break my bones,
but terminology will never hurt me.

There are some similarities, but many differences, between traditional and GIS maps. This section describes the conceptual differences and terminology used in vector and raster map formats as well as an overall organizational structure for GIS databases.

Geographical information system (GIS) technology has its roots in computer mapping and spatial database management. It allows users to effectively organize, update, and query mapped data. More recently, GIS has moved from graphic inventory of the landscape to modeling potential land uses. The evolution from mapping to data management to modeling is the result of the digital map format and increasing quantification of map analysis procedures. The new procedures and resulting decision-making environment require a rethinking of traditional mapping concepts. Notions of error propagation, weighted distance measurement, visual exposure surfaces, nth optimal paths, spatial statistics, and fragmentation indices form some of the new tools confronting the users of GIS.

Coming to Terms with the Terminology

To some, the unfamiliar terminology, concepts, and capabilities of map analysis represent the "darker side" of GIS technology. Traditional mapping and database management are comfortable turf. You use the computer to link your file cabinets to your map sheets; it automates your daily routine. Your current concepts are easily transferred. But the analytic capabilities of GIS takes us well beyond mapping. It challenges old assumptions and suggests new applications. In many respects, GIS is more different than it is similar to traditional map processing. The most obvious changes are in the GIS maps themselves. New terms and concepts abound. To go beyond mapping, you must first become comfortable with the basic GIS terminology and organizational structure. This introduction is designed to develop this foundation.

Like other new technologies, GIS is often guilty of concealing its basic concepts in unfamiliar terminology. The concepts are simple; it's the terms that are complex. Once you cut through the hyperbole, GIS is a lot like what you do now. In the real world, the landscape is composed of rocks, dirt, trees, and fine feathered friends. In your "paper" world, these things are represented by words, tables, and graphics.

Your maps are graphical abstractions in which inked lines, shadings, and symbols are used to locate landscape *features*. Technically speaking, all maps are composed of three basic features: *points, lines,* and *areas*. For example, a typical water map identifies a spring as a dot, a stream as a squiggle, and a lake as a blue glob. GIS can reproduce a similar graphic, but that isn't how it stores the data. In the GIS world, map features most often are represented by *x,y coordinates*, as shown in figure I.1. Points are identified as a single coordinate pair. Lines are identified as a connected set of points (like connect-the-dot pictures). Areas, such as an ownership parcel, are identified by the coordinates defining their borders. This comfortable data format is termed *vector*.

A less familiar format, termed *raster*, uses an imaginary grid of cells to represent the landscape. Points are stored as individual *column,row entries*. Lines are stored as a set of connected cells. Areas are identified as all of the cells within the interior of each feature. Although this data structure has several advantages, it has a major disadvantage—a

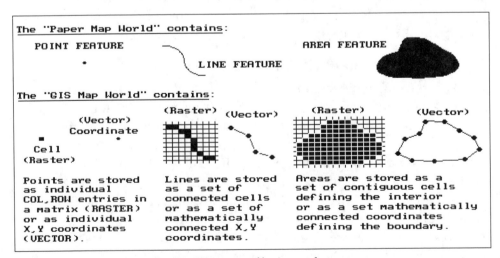

Fig. I.1. GIS storage of basic map features.

lack of precision. If a stream passes through an acre cell, the whole cell is identified as "a stream." You don't know if it is at the top, bottom, or wiggles several times through the center of the cell. Further discussion of data structure is best reserved for later.

For now, let's see how maps are linked to data. In the paper world you are the link, running back and forth between a map and your file cabinets. If you want to know which timber stands have Douglas fir and Cohassett soil, you flip through your files and note the stand numbers of those you're after. You go to the map to locate them. If you wonder what the forest/soil type is for a neighboring stand, you run back to the files to look it up.

That's a lot of work for you, but not for your GIS. The map and file cabinets are electronically linked as shown in figure I.2. A common *identification number* (ID#) is part of the *map features* and the *thematic attribute* tables. Actually, these tables are just plain old databases—one stores the *x,y* coordinates, while the other stores the information about each stand. Each row of the attribute table (termed a *record*) is divided into several columns (termed *items*). This is similar to the old days when foresters kept data (items) on a 4-x-5 card (record) for each stand. In confusing "techy-speak," the items COVER = DF and SOIL = COH are searched. In the figure, stand 4 is the only one that meets the joint condition. Its coordinates are plotted to the screen and filled with a vibrant color of your choice. The GIS searches the remaining 40,000 stands, and all of the "hits" are plotted in less than a minute. If

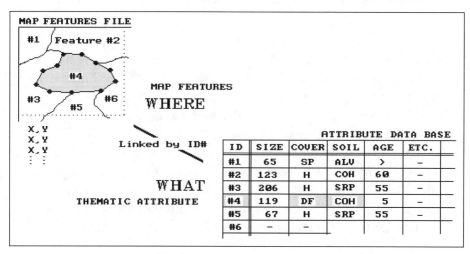

Fig. I.2. Linking map features and attribute data.

you click your "mouse" anywhere on the map, the data about that stand pop up in less than a second. Think of the shoe leather you could save.

The raster world has an analogous link. Each cell has an implicit ID# based on its column, row position. By convention, the analysis grid is ordered as you read a book, from left to right, top to bottom. By implication, the first cell (ID# 1) is in the upper-left corner. The next cell (ID# 2) is the adjacent cell to the right. The sequential numbering continues through the last column of the first row. It then picks up with the first column of the second row and continues the left to right sequence for each successive row. It finally finishes with the lower-right cell. Most raster systems store the *what* information in a separate attribute table for each map.

Some systems store the information as one large table with each record indicating a cell and each item describing a separate thematic attribute. If you think about it (see fig. I.2), the similarities between the vector and raster formats should be apparent. The attribute databases, containing the *what* information, are nearly identical with the exception of the ID#'s—explicit for vector, implicit for raster. The map features file, containing the *where* information, for vector stores irregular features, whereas for raster it is an implied analysis grid of regular cells. There are a lot of similarities between the two but there are also some significant differences.

GIS Maps Are Dumb without a Data Structure

When you view a map, all sorts of things are apparent. If two blue lines come together you instantly recognize it as a fork in a stream. As your eyes moves along a set of blue lines, you easily comprehend which stream networks are connected and which are not. Your interpretation of the contour lines even tells you which way the water is flowing.

The GIS isn't as lucky. With your map view, you see it all and bring to bear years of experience, insight, and intuition. When a GIS "views" a map, it does it a piece at a time. You're holistic, it's myopic. The relationships among the pieces (termed *spatial topology*) has to be contained in the data's organization. The GIS may know that a coordinate pair (*x,y* location) is identified with a stream, but without topology it has no idea how that location relates to all of the other map locations.

The fundamental element of map structure is the *point*, which is represented by a pair of *x,y* values. These coordinates usually relate to a standard referencing grid such as latitude and longitude or Universal Traverse Mercator (UTM). There are several types of points, as shown in figure I.3. *Tics* are geographic control points used in registering a map. *Discrete* points are used to represent information such as wells. Also, they are used to associate data with areas and to position map labels. *Vertices* and *nodes* are used to construct lines and areal features. This process is similar to the connect-the-dots drawings from your childhood. You started with the first dot, then drew from one to the next until things took shape. Vertices are merely passed through, whereas nodes identify the special points where more than two lines meet.

The set of line segments between two nodes is referred to as an *arc*. In the case of linear features, such as a stream network, the GIS keeps track of which arcs are connected. Also it notes the up-, downstream nodes of each arc. When the stream flows into a lake the node is

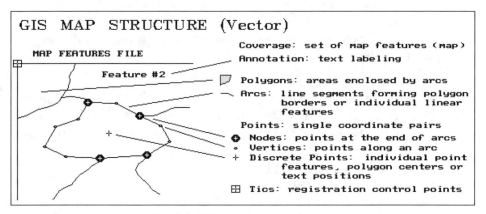

Fig. I.3. GIS map structure (vector).

tagged as an inlet. The stream node at the other end of the lake is identified as an outlet. You see this stuff; the computer has to be told.

Polygons are areas enclosed by arcs. Just as several points form an arc, a closed series of arcs form polygons. There is a discrete point inside each polygon, which serves as a link to the information about it (e.g., size, cover, soil, age). When polygons are adjoining, such as timber stands, their shared arcs are tagged with a special code identifying the linkage. In this manner, the GIS knows the adjoining polygons, and their adjoining polygons, and so on.

With minimal guidance from map *annotations*, you see all this. However, the GIS must incorporate it into its data structure. When it jumps into the middle of a map (termed a *coverage*), it has to be able to sequentially construct all of the relationships among the map features. Yep, GIS maps are dumb. It's a good thing the computer keeps track of all the details.

Figure I.4 shows the same area expressed in raster format. The entire landscape is covered by an imaginary grid of *cells*, the basic unit of this data structure. There are two types of cells. A *whole cell* contains a single map characteristic throughout its interior (e.g., soil or forest type). A *partial cell* contains a mixture of characteristics (e.g., part soils A and B) or just a portion of an individual characteristic (e.g,. road or spring). It's the partial cells that account for the lack of precision of raster data. The entire area of a cell is the smallest addressable unit and all spatial detail smaller than a cell is lost. If a finer analysis grid is used, precision increases. In theory, the grid could be as fine as the x,y coordinates in a vector system, yielding identical precision. However, the storage and processing demands at such a high resolution exceed the capacities of most modern computers.

Until there is a supercomputer on every desk, an oversized partial cell must be used to identify a single point in space. Connected series of partial cells are used to identify *lines*. And, a set of whole (interior)

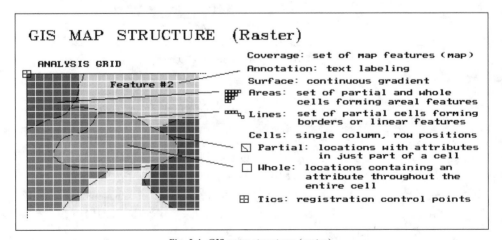

Fig. I.4. GIS map structure (raster).

and partial (border) cells are used to identify *areas*. To simplify things, the characteristic dominating a border cell can be assigned, thus making it a whole cell (and a whole lot easier to store).

The raster structure allows us to extend the basic map features from just points, lines, and areas, to *surfaces*. A surface describes the continuous distribution of gradient data. Elevation is a good example, at least in hilly terrain. Each cell is assigned an elevation value that typifies the elevation within its boundary. The raster format of elevation data is termed a digital elevation model (DEM) and contains radically different information than the traditional contour map. Atmospheric pressure, temperature, and cost surfaces are other examples of this new type of map feature. When you think about it, we have just "scratched the surface" of this strange beast called GIS.

Terminology Accelerates Your Intellectual Depletion Allowance

So far, the basic terminology and approaches in data structure have been discussed. However, before we can jump into the implications of treating maps as data and data structure alternatives, there is a bigger picture that has to be covered, *workspace* organization. Recall that a *coverage* is the proper term for a GIS map (vector or raster format alike). Three things make a GIS coverage different from a traditional map: It's digital; it represents only one theme; and it's *seamless*. Seamless means a user can specify a set of corner coordinates and the GIS will automatically "cut and paste" data from the appropriate storage *sections* (see fig. I.5). This process is similar to your locating four adjoining topographic sheets, identifying your project area boundary on each, whacking away with your scissors, and then taping the pieces together. Like sections in the public land survey system (PLSS), a GIS section is simply a means of dividing a large area into regular blocks for efficient referencing. Most users are unaware of the computer's fundamental organization of sections as the seamless database structure allows them to define any project area they please.

A *layer* is a set of adjoining sections having the same features and attributes. For example, a GIS database might contain separate layers for political jurisdictions, roads, elevation, hydrography, and vegetative cover. In contrast, the familiar U.S. Geological Survey (USGS) topographic map depicts all these plus other themes on a single map

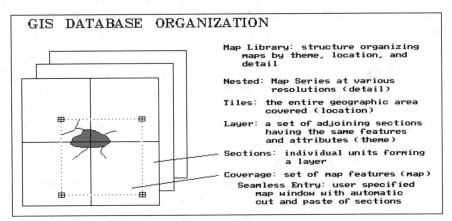

Fig. I.5. GIS database organization.

sheet. That's what you see, but that's not really the case. Actually, each theme is stored separately and printed as a sandwich of inked layers.

It is imperative that great care is taken in encoding each section or they might not *edge match*. As shown in figure I.5, the boundaries of features must continue from section to section. Misalignment of edges is the most frequent cause of premature GIS death. Obviously, your registration and digitizing must be extremely precise, but that may not be enough. A couple of uncontrollable problems can arise. The original maps you're encoding might not align. If so, adjust them the best you can. Or, more subversively, the classification scheme may not be consistent. For example, you might encode two abutting forest maps with one having six levels of stocking/age classes, and the other having eight. No matter how carefully you digitize they will never edge-match (and your GIS is doomed from the start).

A layer describes the informational content of each GIS map. A *tile* describes the basic geographic area represented in each of the layers. Although tiles are generally rectangular, they may be any shape, such as a county or forest administration unit. You can think of them as the digital analogue for the map sheets of a conventional map series.

A *nested* map series contains maps at various resolutions over the same geographic area. This concept is particularly applicable to raster databases containing different satellite data. Frequently, a user will store two or more analysis grid resolutions of the same area—a coarse one for strategic and fine one for tactical studies. The familiar USGS's 7.5- and 15-minute topographic series is a paper product example of nesting.

One final concept ties it all together, the *map library*. A map library refers to a listing of all GIS maps in a system. The listing is simultaneously organized by location, theme, and detail. In a full-featured GIS, you can specify a project area, select the maps you need, then store them in your own workspace. With a healthy understanding of the errors introduced, you can transform maps of various geographic scales and projections, resample maps of various levels of detail, as well as exchange vector and raster maps.

Whew! All this has been an overload in both mundane and arcane terminology. Some of it makes common sense; some of it may make no sense at all. Keep in mind that you're the intellectual superior of the GIS. You simply see things, while it has to organize everything in excruciating detail. Although fundamentally different, you and your GIS need to be agreeable partners. Table I.1 summarizes all the terms in this "partnership."

Table I.1. Spatial data terms.

Data Characterization
Column, row entries
Features
Identification number
Items
Map features
Points, lines, areas
Raster
Records
Thematic attribute
Vector
x,y coordinates

Vector Data Model
Annotation
Arcs
Coverage
Discrete points
Nodes
Point
Polygons
Tics
Topology
Vertices

Raster Data Model
Area cell set
Cell (column, row)
Line cell series
Partial cell
Surfaces
Whole cell

Data Organization
Edge-match
Layers
Map library
Nested
Seamless
Sections
Tiles
Workspace

Recommended Reading

Books

Aronoff, S. "Data Management." Chapt. 6 in *Geographic Information Systems: A Management Perspective*. Ottawa, Canada: WDL Publications, 1989.

Burrough, P.A. "Geographical Information Systems." Chapt. 1 in *Principles of Geographical Information Systems for Land Resources Assessment*. Oxford, UK: Oxford University Press, 1987.

Dangermond, J. *Software Components Commonly Used in Geographic Information Systems*. Redlands, CA: Environmental Systems Research Institute, 1983.

Franklin, W.R. "Computer Systems and Low-Level Data Structures for GIS." In *Geographical Information Systems: Principles and Applications*, ed. D.J. Maguire, M.F. Goodchild, and D.W. Rhind, Vol. 1, 215-25. Essex, UK: Longman, 1991.

Maling, D.H. "Coordinate Systems and Map Projections for GIS." In *Geographical Information Systems: Principles and Applications*, ed. D.J. Maguire, M.F. Goodchild, and D.W. Rhind, Vol. 1, 34-46. Essex, UK: Longman, 1991.

Morehouse, S. *Arc/Info: A Geo-Relational Model for Spatial Information*. Redlands, CA: Environmental Systems Research Institute, 1990.

Parent, P., and R. Church "Evolution of Geographic Information Systems as Decision Making Tools." In *Fundamentals of Geographic Information Systems: A Compendium*, ed. W. Ripple, 9-18. Bethesda MD: American Society of Photogrammetry and Remote Sensing, 1989.

Shepherd, I.D. "Information Integration and GIS." In *Geographical Information Systems: Principles and Applications*, ed. D.J. Maguire, M.F. Goodchild, and D.W. Rhind, Vol. 1, 337-60. Essex, UK: Longman, 1991.

Star, J., and J. Estes. "Introduction," "Background and History," and "The Essential Elements of a GIS: An Overview." Chapts. 1, 2, and 3 in *Geographic Information Systems: An Introduction*, Englewood Cliffs, NJ: Prentice Hall, 1990.

Thompson, D. "Geographic Information Systems: An Introduction." *1991-92 International GIS Sourcebook*, 338-41. Fort Collins, CO: GIS World, 1991.

Journal Articles

Berry, J.K. "Computer-Assisted Map Analysis: Potential and Pitfalls." *Photogrammetric Engineering and Remote Sensing*, 53(10): 1405-10 (1987).

Devine, H. "GIS In the 90's." *Compiler* 8(3): 5-14 (1990).

Piwowar, J.M., et. al. "Integration of Spatial Data in Vector and Raster Formats in a Geographic Information System." *International Journal of Geographical Information Systems* 4(4): 429-44 (1990).

Pueker, T., and N. Christman. "Cartographic Data Structures." *American Cartographer* 2(1): 55-69 (1990).

Waugh, T., and R. Healey. "A Relational Data Base Approach to Geographical Data Handling." *International Journal of Geographical Information Systems* 1(2): 101-18 (1987).

TOPIC 1

MAPS AS DATA AND DATA STRUCTURE IMPLICATIONS

Our brains evolved to get us out of the rain, find where berries are, and keep us from getting killed. Our brains did not evolve to help us grasp really large numbers or look at things in a hundred thousand dimensions.

—Ronald Grahm, Mathematician, Bell Labs

The full impact of numerical representation of spatial data in GIS is just beginning to be recognized. In this section the implications of the vector and raster data models on encoding, storage, and analysis are discussed. Also, the inherent statistical characteristics of mapped data and their implications in map analysis are described.

1 Maps as Data: An Emerging "Map-ematics"

(Investigating the Digital Nature of Maps)

❖   ❖   ❖

Old proverb: A picture is worth a thousand words.
New proverb: A map is worth a thousand numbers,
maybe more.

Maps and Mapping

Our historical perspective of maps is one of accurate location of physical features primarily for travel through unfamiliar areas. Early explorers used them to avoid angry serpents, alluring sirens, and even the edge of the earth. The mapping process evokes images of map sheets and drafting aids such as pens, rub-on shading, rulers, planimeters, dot grids, and acetate transparencies for light-table overlays—sort of a Keystone Cops comedy of cartographic processing. From this perspective, maps are analog mediums composed of lines, colors, and symbols that are manually created and analyzed. Because manual analysis is difficult and limited, the focus of the analog map and manual processing has been descriptive—recording the occurrence and distribution of landscape features.

More recently, the analysis of mapped data has become an integral part of resource and land planning. By the 1960s, manual procedures for overlaying maps were common. These techniques marked a turning point in the use of maps, from techniques that emphasize the physical descriptors of geographic space to those that spatially characterize management actions. This movement from descriptive to prescriptive mapping set the stage for computer-assisted map analysis.

Since the 1960s all aspects of decision making have become much more quantitative. Mathematical models for nonspatial analyses are now commonplace. However, the tremendous volume of data used in spatial analysis limits the application of traditional statistics and mathematics in spatial modeling. Nonspatial procedures require that maps be generalized to typical values before they can be used. Thus, the spatial detail for large areas is often reduced to a single value expressing the central tendency of a variable over that area. This results in a tremendous reduction in information from the spatial specificity in the original map—a real bummer if you are a stickler for details.

Recognition of this problem led to the stratification of regions at the beginning of a study by dividing geographic space into assumed homogeneous sampling parcels. Heated debates often arise as to whether a normal, binomial or Poisson distribution best characterizes the typical value in numeric space. However, relatively little attention is given to the broad assumption that this value must be presumed to be uniformly distributed in geographic space. The area-weighted average of several parcels' typical values frequently is used to characterize an entire study area. Mathematical modeling of spatial systems has followed an approach similar to that of spatial statistics, aggregating the spatial variation of model variables. Most ecosystem models, for example, define "level" and "flow" variables that are presumed to be typical for vast geographic expanses.

However, maps actually map the details of spatial variation. Manual cartographic techniques allow manipulation of these detailed data, yet they are fundamentally limited by their nondigital nature. Traditional statistics and mathematics are digital, yet they are fundamentally limited by their generalization of the data. Such was the dilemma a decade ago. This dichotomy has led to the revolutionary concepts of map structure, content, and use that form the foundation of GIS technology. It radically changes our perspective. Maps move from analog images describing the distribution of features to geographically referenced digital data quantifying a physical, social, or economic system in prescriptive terms.

Digitizing Maps and Mapping

This revolution is founded in the recognition of the digital nature of computerized maps—maps as data, maps as numbers. To illustrate, consider figure 1.1. The upper-left inset is a typical topographic map. One-hundred-foot contour lines show the pattern of the elevation gradient over the area. The human eye quickly assesses the flat areas, the steep areas, the peaks, and the depressions. However, in this form the elevation information is incompatible with any quantitative model requiring input of this variable. Descriptive statistics can be used to generalize the elevation gradient as shown in the table in the upper-right. We note that the elevation ranges from 500 to 2,500 feet with an average of 1,293 feet. The standard deviation of ±595 feet tells us how typical this average is; most often (about two-thirds of the time), we would expect to encounter elevations from 698 to 1,888 feet. But where would one expect higher or lower elevations? The statistic offers no insight other than the larger the variation, the less typical is the average—the smaller the better. In this instance, it's not very good because the standard deviation is nearly half the mean (coefficient of variation = .46).

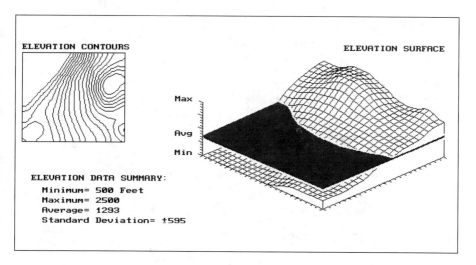

Fig. 1.1. Conventional elevation (topographic) contour map versus three-dimensional terrain representation.

The larger centered inset in figure 1.1 is a three-dimensional plot of the elevation data. The gridded data contain an estimate of the elevation at each hectare throughout the area. In this form both your eye and the computer see the variability in the terrain, the flat area in the northwest, the highlands in the northeast. For contrast, the average elevation is represented as the horizontal plane intersecting the surface at 1,293 feet. Its standard deviation can be conceptualized as two additional planes floating ±595 feet above and below the average plane. A nonspatial model that must assume the actual elevation for any parcel is somewhere between these variation planes, most likely about 1,293 feet.

But your eye notes that the eastern portion is above the mean, while the western portion is below. The digital representation stored in a GIS tracks this variation in quantitative terms. Thus, the average and variation provide conceptual linkages between spatial and nonspatial data. The regional average used in traditional statistics reduces the complexity of geographic space to a single value. Spatial statistics retain this complexity by depicting it on a map of the variation in the data. Thus, a regionalized plane represents the mean and is contrasted with the continuously varying surface that represents the variance.

In computer-assisted map analysis all maps are viewed as a set of ordered numbers. These numbers have numerical significance as well as conventional spatial positioning information. It is the numerical attribute of GIS maps that fuels the concept of "map-ematics." For example, the first derivative of the elevation surface in figure 1.1 creates a slope map. The second derivative creates a terrain roughness map (where slope is changing). An aspect map (azimuthal orientation) indicates the direction of terrain slope at each hectare parcel. But what if the figure wasn't mapping elevation rather the concentration of an environmental variable, such as lake temperatures or soil concentrations of lead? For lake temperatures, the first derivative would map the rate of temperature change. The aspect map would indicate the direction of change throughout the lake. For lead concentrations, the first derivative would map the rate of change in concentration, and the second derivative (change in the rate of change in concentration) could provide information about multiple sources of lead pollution or abrupt changes in seasonal wind patterns. The aspect map of lead concentrations would indicate the direction of change in concentration.

All this seems terribly complex. Maps as data, map-ematics, a derivative of a map! These concepts might at first seem strange, but as you become more familiar with the digital nature of maps in a GIS, they will become as comfortable as a pair of old woolen socks. Truly, a map is indeed worth a thousand numbers—often more.

2 | "It Depends": Implications of Data Structure

(The Relative Merits and Demerits of Alternative Structures)

The main purpose of a geographic information system is to process spatial information. In doing so it must be capable of four things:

- Creation of digital abstractions of the landscape (encode),

- Efficient handling of these data (store),

- Development of new insights into the relationships of spatial variables (analyze), and

- Creation of "human-compatible" summaries of these relationships (display).

The data structure used for storage has far-reaching implications in how we encode, analyze, and display digital maps. The question has fueled heated debate as to the "universal truth" in data structure since the inception of GIS. In truth, there are more similarities than differences in the various approaches.

As discussed in the previous article, all GISs are internally referenced, which means they have an automated linkage between the data (or thematic attributes) and the whereabouts (locational attributes) of those data. There are two basic approaches used in describing locational attributes. One approach (vector) uses a collection of line segments to identify the boundaries of point, linear, and areal features. The alternative approach (raster) establishes an imaginary grid pattern over a study area, then stores values identifying the map characteristic occurring within each grid space. Although there are significant practical differences in these data structures, the primary theoretical difference is that the grid structure stores information on the interior of areal features, and implies boundaries, whereas the line structure stores information about boundaries, and implies interiors. This fundamental difference determines, for the most part, the types of applications that may be addressed by a particular GIS.

X's, Y's; Columns and Rows

It is important to note that both systems are actually grid based. It's just in practice that line-oriented systems use a very fine grid of digitizer coordinates. From the computer's perspective point features, such as springs or wells on a water map, are stored the same for both systems—a single digitizer "x,y" coordinate pair or a single "column, row" cell identifier. Similarly, line features, such as streams on a water map, are stored the same—a series of "x,y" or "column, row" identifiers. If the same gridding resolution is used, there is no theoretical difference between the two referencing schemes and, considering modern storage devices, only minimal practical differences in storage requirements. Yet, it was storage considerations that fueled most of the early debate about the relative merits of the two data structures. The demands of a few, or even one, megabyte of storage were very important in the 1970s. To reduce storage, very coarse grids were used in early grid systems. With this practice, streams were no longer the familar thin lines assumed a few feet in width but represented as a string of cells of several acres each. This, coupled with the heavy reliance on pen-plotter output, resulted in "ugly, saw-toothed" map products when using grid systems. Recognition of any redeeming qualities of this data form were lost to the unfamilar character of the map product.

Is a Pie Filling Or Crust?

Consideration of areal features presents significant theoretical differences between the two data structures. A lake on a water map may be described by its border defined as a series of line segments, or its interior defined by a set of cells identifying open water. This difference has important implications in the assumptions about mapped data. In a line-based system, the lines are assumed to be "real" divisions of geographic space into homogenous units. This assumption is reasonable for most lakes if you accept the premise that the shoreline remains constant. However, if the body of water is a flood-control reservoir, the shoreline could shift several hundred meters during a single year; a "fat, fuzzy" line would be more realistic. A better example of an ideal line feature is a property boundary. Although these divisions are not physical, they are real and represent indisputable boundaries. One footstep over the line can jeopardize both friendships and international treaties.

However, consider the familar contour map of elevation. The successive contour lines form a series of long skinny polygons. Within each of these polygons the elevation is assumed to be constant—forming a "layer-cake" of flat terraces in three-dimensional data space. For a few places in the world, such as rice patties in mountainous portions of Korea or the mesas of New Mexico, this may be an accurate portrayal. This aggregation of a continuous spatial gradient discards much of the information used in its derivation. An even less clear example of a traditional line-based image is the familar soil map. The careful use of a fine-tipped pen in characterizing the distribution of soils imparts artificial accuracy at best. At worst, it severely limits the potential uses of soil information in a GIS. As with most resource and environmental data, a soil map is not "certain," as contrasted with a surveyed and legally filed property map. Rather the distribution of soils are probabilistic, but the lines form artificial boundaries presumed to be the abrupt transition from one soil to another. Throughout each soil polygon, the occurrence of the designated soil is treated as equally likely. Most users of soil maps reluctantly accept the "inviolately accurate" assumption of this data structure; the alternative is to dig soil pits everywhere within a study area. It's a lot easier to just go with the flow.

A more useful data structure for representing soils is gridded, with each grid location identified by its most probable soil, a statistic indicating how probable, the next most probable soil, its likelihood, and so on. In this context, soils are characterized as statistically continuous gradients rather than as aggregated, human-compatible images (see topic 6). Such treatment of map information is a radical departure from the traditional cartographic image. It highlights the revolution in spatial information handling brought about by the digital map.

Of Polygons and Cells

All maps generalize detail over some geographic unit. When do collections of molecules, particles, dirt clods, piles, or acres become the units on a soil map? What variablity in composition is tolerated? In GIS, this assumed uniform unit is termed a *point*. It is defined as the smallest addressable unit of space and should not be confused with geometry's definition of a point as "having neither length nor width." GIS points are just the opposite—the smallest area identifying a map characteristic.

Sets of points form "regions" on a map. For example, a map of WATER (i.e., coverage in GIS jargon) is comprised of several different water types (i.e., regions), such as lakes, streams, and wetlands. Each lake, stream, or wetland is identified as a set of points stored in the GIS —polygons in the line-based structure or cells in the grid-based structure. It is important to note that both systems are actually polygon based; it is just in practice that most grid-based systems utilize a regular grid of square polygons (cells). Vector systems could store coordinates identifying the four sides of each cell forming an array of contiguous, uniform polygons. If these polygons are the same size as cells in raster system, there is no theoretical difference between the two referencing schemes and no practical advantage in locational accuracy.

Yet, like storage, spatial precision continued to fuel early debate about the "best" data structure. Course cells did not portray the apparent accuracy in the cuts and jogs of the thin blue line representing a stream on a USGS topographic map. Your experience, however, may have led you to question this implied accuracy. The same stream is often characterized by the same pen width in both 7 1/2- and 15-minute series of maps—does the stream change width or just mapping scale? Similarly, if you enlarge the smaller scale map and overlay the two maps, you will notice that many of the cuts and jogs have been "smoothed."

"Appropriate" Structure Is "Best"

It's safe to say that there isn't an absolute "best" data structure. It depends. Contemporary hardware and software technology play important roles. However, *absolutes*, defined by perceived practical limits, are continually shattered. Not so long ago the functions of today's 26-key pocket calculator were the domain of the mainframe computer.

A more enduring comparision of the two data structures is based on their theoretical differences and resulting implications on encoding, storage, analysis, and display. Consider table 2.1 outlining some of these factors. The dichotomy drawn is the result of the nature of the mapped data and the intended application.

If you are a fire chief, your data consist of surveyed roads, installed fire hydrants, and tagged street addresses (lines real and data certain).

Whenever a fire is reported, you query the database to map the route to the fire and report the closest fire hydrants (descriptive application involving computer mapping and spatial database management).

Table 2.1. Comparison of two data structures and their applications.

Vector (Inventory)	Raster (Analysis)
• Lines real	• Lines artificial
• Data known	• Data probabilistic
• Descriptive queries	• Prescriptive analysis
• Computer mapping	• Spatial statistics
• Spatial DBMS	• Modeling

Contrast this use with that of a resource manager working with maps of soils, vegetation, and wildlife habitat (lines artificial and data probabilistic) to identify the optimal timber harvesting schedule for an area (prescriptive application involving spatial statistics, and modeling). You know, different strokes for different folks.

Generally speaking, the fire chief's application is best solved with a vector-based data structure considering precise locations with minimal analysis. Think of the consequences of a raster data structure in such an application—a .85 probability that a fire hydrant is located in a one-block grid cell. Resource managers, on the other hand, have an entirely different set of problems. Their applications, involving complex analyses and minimal spatial precision are best met with a raster structure. Think of the consequences of a vector data structure— absolute certainty that a quarter-acre parcel is ponderosa pine on Cohassett soil. Both results are equally inappropriate. The "best" data structure depends on the nature of the data used in the analysis and the analytic tools required by the application.

3 GIS Technology Is Technical Oz

(Data Structure Effects on Map Analysis Capabilities with GIS)

> You're hit with a tornado of new concepts, temporarily hallucinate, and come back to yourself a short time later wondering what on earth all those crazy things meant.
>
> *–J.K. Berry*

Recall that, first and foremost, maps in a GIS are digital data organized as large sets of numbers, not analog images comprised of inked lines, colors, and shadings. Data structure refers to how we organize these numbers—basically as a collection of line segments or as a set of grid cells. Theoretical differences between these two structures arise for storage of areal features. Line-based structures store information about area boundaries and imply interiors. Cell-based structures do just the opposite, implying boundaries while storing information on interiors. So much for review. What does this imply for map analysis?

Vector vs. Cell Structures in Map Analysis

In short, which of the two basic approaches is used in processing significantly affects map analysis speed, accuracy, and storage requirements. Structure also defines the set of potential analytic "tools" and their algorithms. For example, consider figure 3.1 depicting three simple geometric shapes stored in typical formats of both structures. The vectors describe boundaries around areas assumed to be the same throughout their interiors (fig. 3.1a). Cells, on the other hand, define the interior of features as groupings of contiguous cells (fig. 3.1b). The spatial precision of the boundaries is obviously better for the line structure. The sawtooth effect in the grid structure is an unreal and undesirable artifact. It's fair to say that the line structure frequently has an advantage in spatial preci-

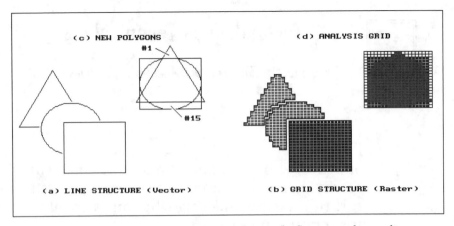

Fig. 3.1. Comparison of vector (a, c) and raster (b, d) map overlay results.

sion and storage efficiency of base maps. Therefore, it is generally better for inventory.

Other differences are apparent during analysis of these data. For example, the composite maps (fig. 3.1c, d) are the result of simply overlaying the three features, one of the basic analytic functions. In the line structure (fig. 3.1c), 15 new polygons are derived, composed of 36 individual line segments. This is a significant increase in the storage requirement for the composite map as compared to any of the simple original maps. But consider the realistic complexity of overlaying a land-use map of several hundred polygons with a soil map of over a thousand. The result is more "son and daughter" polygonal progeny than you would care to count (or most small computers would care to store). On the other hand, the storage requirement for the grid structure (fig. 3.1d) can never exceed the maximum dimensionality of the grid, no matter how many input maps or their complexity. Even more significant are the computational demands involved in splitting and fusing the potentially thousands of line segments forming the new boundaries of the derived map. By contrast, the overlaying of the maps stored in grid structure simply involves direct storage access and matrix addition. It's fair to say that the grid structure frequently has an advantage in computation and storage efficiency of derived maps. Therefore, it is generally better for analysis.

Also, it is fair to say that the relative advantages and disadvantages of the two data structures have not escaped GIS technologists. Database suppliers determine the best format for each variable. (For example, the

U.S. Geological Survey uses vector format for all 7.5 minute quadrangle information except elevation, which it supplies in raster format). Most vendors provide conversion routines for transferring data between vector and raster. Many provide "schizophrenic" systems with both a vector and a raster processing side. Some have developed specialized data structure offshoots, such as rasterized lines, quadtrees, and triangular irregular networks (TIN). In each instance, careful consideration is made to the nature of the data, processing considerations, and the intended use—it depends.

Data Characteristics in Map Analysis

Another concern is the data characteristics derived in map analysis. In the case of line structure, each derived polygon is assumed to be accurately defined: precise intersection of real boundaries surrounding a uniform geographic distribution of data. This is true for overlaying a property map with a zip code map, but a limiting assumption for probabalistic resource data, such as soils and land cover, as well as gradient data, such as topographic relief and weather maps. For example, consider the geographic search (overlay) for areas of Cohassett soil, moderate slope, and ponderosa pine forest cover. A line-based system generates an "image" of the intersections of the specified polygons. Each derived polygon is assumed to locate the precisely defined combinations of the variables. In addition, the likelihood of actual occurrence is assumed to be the same for all of the polygonal progeny, even small *slivers*, areas formed by intersecting edges of the input polygons.

A grid-oriented system calculates the coincidence of variables at each cell location as if each were an individual polygon. Because these polygons are organized as a consistent uniform grid, the calculations simply involve storage retrieval and numeric evaluation, not geometric calculations for intersecting lines. In addition, if an estimate of error is available for each variable at each cell, the value assigned as a function of these data also can indicate the most likely composition (coincidence) of the variables: "There is an 80-percent chance that this hectare is Cohassett soil, moderately sloped, and ponderosa pine covered." The result is a digital map of the derived variable, expressed as a geographic distribution, plus its likelihood of error (a sort of "shadow" map of certainty of result). This concept, termed error propagation modeling (see topic 6), is admittedly an unfamiliar and, likely, an uncomfortable one.

It is but one of the gusts in the GIS whirlwind that is taking us beyond mapping. Others include drastically modified techniques, such as weighted distance measurement (a sort of "rubber ruler"), and entirely new procedures, such as optimal path density analysis (identifying the nth best route). These new analytic concepts and constructs are the focus of topic 2.

Recommended Reading

Books

Burrough, P.A. "Data Structures for Thematic Maps." Chapt. 2 in *Principles of Geographical Information Systems for Land Resources Assessment*. Oxford, UK: Oxford University Press, 1987.

Dangermond, J. "GIS Data Structures: Objects vs. Layers." *1989 International GIS Sourcebook*,18-20. Fort Collins, CO: GIS World, 1989.

Egenhofer, M., and J. Herring. "High-Level Spatial Data Structures for GIS." In *Geographical Information Systems: Principles and Applications*, ed. D.J. Maguire, M.F. Goodchild, and D.W. Rhind, Vol. 1, 227-37. Essex, UK: Longman, 1991.

Piazza, P., and F. Pessaro. "A Cognitive Model for a 'Smart' GIS." *1990 International GIS Sourcebook*, 273-79. Fort Collins, CO: GIS World, 1990.

Raper, J., and B. Kelk. "Three-Dimensional GIS." In *Geographical Information Systems: Principles and Applications*, ed. D.J. Maguire, M.F. Goodchild, and D.W. Rhind, Vol. 1, 299-317. Essex, UK: Longman, 1991.

Star, J., and J. Estes. "Data Structures" and "Data Management." Chapts. 4 and 7 in *Geographic Information Systems: An Introduction*, Englewood Cliffs, NJ: Prentice Hall, 1990.

Tomlin, C.D. "Data." Chapt. 1 in *Geographic Information Systems and Cartographic Modeling*. Englewood Cliffs, NJ: Prentice Hall, 1990.

Journal Articles

Bolstad, P., et al. "Uncertainty in Manually Digitized Mapped Data." *International Journal of Geographical Information Systems* 4(4): 399-412 (1990).

Folse, L.J., et al. "Object-Oriented Simulation and Geographic Information Systems." *AI Applications* 4(2): 41-47 (1990).

Frank, A.U. "Requirements for a Database Management System for a GIS." *Photogrammetric Engineering and Remote Sensing* 54(11): 1557-64 (1988).

Maffini, G. "Raster Versus Vector Data Encoding and Handling: A Commentary." *Photogrammetric Engineering and Remote Sensing*, 53(10): 1397-98 (1987).
Van Rossel, J.W. "Design of a Spatial Data Structure Using Relational Normal Forms." *International Journal of Geographical Information Systems* 1(1): 33-50 (1987).

TOPIC 2

MEASURING EFFECTIVE
DISTANCE AND CONNECTIVITY

Make things as simple as possible, but no more so.
–Einstein

Before GIS technology, the concept of distance was as simple and straightforward as a ruler. In this section the traditional concept of distance is first extended to one of proximity, then to one of actual movement in geographic space, around and through barriers. The procedures and applications of optimal path analysis over continuous map surfaces also are presented.

You Can't Get There from Here

(Measuring Simple Distance and Proximity)

❖   ❖   ❖

Measuring distance is one of the most basic map analysis techniques. However, the effective integration of distance considerations in spatial decisions has been limited. Historically, *distance* is defined as "the shortest straight line between two points." While this measure is both easily conceptualized and implemented with a ruler, it is frequently insufficient for decision making. A straight-line route may indicate the distance "as the crow flies," but offer little information for the walking crow.

The Fastest Is Not Always the Shortest Distance

It is equally important to most travelers to have the measurement of distance expressed in more relevant terms, such as time or cost. Consider the trip to the airport from your hotel. You could take a ruler and measure the map distance, then use the map scale to compute the length of a straight-line route, say 12 miles. But if you intend to travel by car, it is likely longer. So you use a sheet of paper to form a series of tic marks along its edge following the zigs and zags of a prominent road route. The total length of the marks multiplied times the map scale is the nonstraight distance, say 18 miles. But your real concern is when shall I leave to catch a nine-o'clock plane, and what route is the best?

Chances are you will disregard both distance measurements and phone the bellhop for advice—24 miles by his back-road route, but you will save 10 minutes. Most decision making involving distance follows this scenario of casting aside the map analysis tool and relying on experience. This procedure is sufficient as long as your experience set is robust and the question is not too complex. The limitation of a map analysis approach is not so much in the concept of distance measurement but in its implementation. Any measurement system requires two components: a standard unit and a procedure for measurement. Using a ruler, the *unit* is the smallest hatching along its edge and the *procedure* is the shortest line along the straightedge. In effect, the ruler represents just one row of a grid implied to cover the entire map. You just position the grid such that it aligns with the two points you want measured and count the grid spaces. To measure another distance you merely realign the grid and count again.

The Shortest Is Not Always the Straightest

The approach used by most GISs has a similar foundation. The *unit* is termed a grid space implied by superimposing an imaginary grid over an area, just as the ruler implied such a grid. The *procedure* for measuring distance from any location to another involves counting the number of intervening grid spaces and multiplying by the *map scale*, termed the shortest straight line. However, the procedure is different because the grid is fixed, so it is not always as easy as counting spaces along a row.

Any point-to-point distance in the grid can be calculated as the hypotenuse of a right triangle formed by the grid's columns and rows. Yet, even this procedure is often too limited in both its computer implementation and information content. Computers detest computing squares and square roots. Because the Pythagorean theorem, just noted, is full of them, many GIS packages use another procedure, *proximity*. Rather than sequentially computing the distance between pairs of locations, concentric equidistance zones are established around a location or set of locations. This procedure is analogous to nailing one end of a ruler at one point and spinning it around. The result is similar to the wave pattern generated when a rock is thrown into a still pond. Each ring indicates one unit "farther away," increasing distance as the wave moves away.

A more complex proximity map would be generated if, for example, all locations with houses were simultaneously considered target locations, in effect, throwing a handful of rocks into the pond. Rings grow until wavefronts meet, then they stop. The result is a map indicating the shortest straight-line distance to the nearest target area (house) for each nontarget area.

In many applications, however, the shortest route between two locations may not be a straight line. And even if it is straight, its geographic length may not always reflect a meaningful measure of distance. Rather, *distance* in these applications is best defined in terms of "movement" expressed as travel time, cost, or energy that may be consumed at rates that vary over time and space.

Distance-modifying effects are termed barriers, a concept implying that the ease of movement in space is not always constant. A shortest route respecting these barriers may be a twisted path around and through the barriers. The GIS database allows the user to locate and calibrate the barriers. The wavelike analytic procedure allows the computer to keep track of the complex interactions of the waves and the barriers.

Two types of barriers are identified by their effects—*absolute* and *relative*. Absolute barriers are those completely restricting movement and therefore constitute an infinite distance between the points they separate. A river might be regarded as an absolute barrier to a non-swimmer. To a swimmer or a boater, however, the same river might be regarded as a relative barrier. Relative barriers are those that are passable but only at a cost that may be equated with an increase in geographical distance; it takes five times longer to row 100 meters than to walk that same distance.

In the conceptual example of tossing a rock into a pond, the waves crash against a jetty in the pond—an absolute barrier the waves must circumvent to get to the other side of the jetty. An oil slick characterizes a relative barrier—waves may proceed through but at a reduced intensity (higher cost of movement over the same grid space). The waves will proceed around and through the oil slick; the one reaching the other side identifies the "shortest, not necessarily straightest line." In effect this is the bellhop's wisdom—he tried many routes to construct his experience base. This same approach is used in GIS, yet the computer is used to simulate these varied paths.

In a GIS, our limited concept of distance as the shortest straight line between two points is first expanded to one of proximity, then to a more effective one of movement through a realistic space containing various barriers. In the past our only recourse for effective distance measurement in "real" space was experience, but deep in your subconcious you know there has to be a better way.

5 As the Crow Walks

(Determining Effective Distance and Optimal Paths)

❖　　❖　　❖

Traditional mapping is in triage. We need to discard some of the old, ineffective procedures and apply new, life-giving technology to others.

–J.K. Berry

The previous discussion of distance measurement with a GIS challenged our fundamental definition of *distance* as "the shortest straight line between two points." It left intact the concept of *shortest*, but relaxed the assumptions that it involves only *two points* and has to be *straight*. In so doing, it first expanded the concept of distance to one of proximity. That is, the shortest, straight line from a location, or set of locations, to all other locations, such as a proximity-to-housing map indicating the distance to the nearest house for every location in a project area. *Proximity* then was expanded to the concept of movement by introducing barriers; the shortest path was not necessarily a straight path. For example, a weighted proximity-to-housing map recognizing the various road and water conditions' effect on the movement of some creatures (flightless, non-swimming crawlers, like us when the car is in the shop).

Effective Proximity

Basic to this expanded view of distance is conceptualizing the measurement process as waves radiating from a location(s), analogous to the ripples caused by tossing a rock into a pond. As the wavefront moves through space, it first checks to see if a potential "step" is passable (absolute barrier locations are not). If it is, it moves to the location and incurs the "cost" of such a movement (relative barrier weights of impedance). As the wavefront proceeds, all possible paths are considered and the shortest distance assigned (least total impedance from the starting point).

The concept is similar to that of a macho guy swaggering across a rain-soaked parking lot as fast as possible. Each time a puddle is encountered, a decision must be reached. Should he proceed slowly so as not to slip, or should he continue a swift, macho pace around the puddle? This distance-related question is answered by experience, not by detailed analysis. "Of all the puddles I have encountered in my life," he mutters, "this looks like one I can handle."

A GIS approaches the question in a much more methodical manner. As the distance wavefront confronts the puddle, it effectively splits. One wave proceeds through at a slower rate and the other goes around at a faster rate. Whichever wave gets to the other side first determines the shortest distance, whether straight or not. The losing wavefront then is totally forgotten and is no longer considered in subsequent distance measurements.

As the wavefront moves through space, it is effectively evaluating all possible paths, retaining only the shortest. You can calibrate a road map so that offroad areas reflect absolute barriers and different types of roads identify relative ease of movement. Then, it is possible to start the computer at a location, asking it to move outward with respect to this complex friction map. The result is a map indicating travel time from the beginning location to every point along the road network. This is the shortest time. Or you can identify a set of starting points, such as a town's four fire houses and have them simultaneously move outward until their wavefronts meet. The result is a map of travel time to the nearest fire house for every location along the road network.

But such effective distance measurement is not restricted to line networks. Take it a step further by calibrating offroad travel in terms of four-wheel, pumper-truck capabilities based on land cover and terrain conditions. Gently sloping meadows are fastest, steep forests are much slower, and large streams and cliffs are prohibitive (an infinitely long time). Identify a forest district's fire headquarters, then move outward respecting both on- and off-road movement for a fire response surface. The resulting surface indicates the expected time of arrival to a fire anywhere in the district.

The idea of a *surface* is basic to understanding both effective distance computation and application. The top left portion of figure 5.1 develops this concept for a simple proximity surface. The tic marks along the ruler identify equal geographic steps from one point to another. If it were replaced with a drafting compass with its point stuck at the lower left, a series of concentric rings could be drawn at each ruler tic mark. This is effectively what the computer generates by sending out a wave-front through unimpeded space. The less than perfect circles in the middle inset of figure 5.1 are the result of the relatively coarse analysis grid used and of approximating errors of the algorithm. They are good estimates of distance but not perfect.

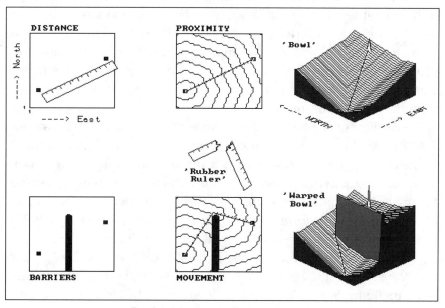

Fig. 5.1. Measuring effective distance.

Accumulation

The real difference is in the information content—less spatial precision but more utility for many applications. A three-dimensional plot of simple distance forms the bowl-like surface in the upper right of figure 5.1. It is sort of like a football stadium with the tiers of seats indicating distance to the field. It doesn't matter which section you're in; if you are in row 100, you had better bring the binoculars. The x- and y- axes determine location while the constantly increasing z-axis (stadium row number) indicates distance from the starting point. If there were several starting points, the surface would be pockmarked with craters, with the ridges between craters indicating the locations equidistant between starters.

The lower-left portion of figure 5.1 shows the effect of introducing an absolute barrier to movement. The wavefront moves outward until it encounters the barrier, then stops. Only those wavefronts that circumvent the barrier are allowed to proceed to the other side, forming a sort of spiral staircase (the lower-middle portion in fig. 5.1). In effect, distance is being measured by a "rubber ruler" that has to bend around the barrier. If relative barriers are present, an even more unusual effect is noted—stretching and compressing the rubber ruler. As the wavefront encounters areas of increased impedance, such as a steep forested area in the fire response example, it is allowed to proceed but at increased time to cross a given unit of space. This has the effect of compressing the ruler's tic marks, which do not represent geographic scale in units of feet but rather the effect on pumper-truck movement measured in units of time.

Regardless of the nature of the barriers present, the result always is a bowl-like surface of distance, termed an *accumulation surface*. Distance always is increasing as you move away from a starter location, forming a perfect bowl when no barriers are present. If barriers are present, the rate of accumulation varies with location, and a complex, warped bowl is formed (the lower-right portion in fig. 5.1). But it is a bowl nonetheless, with its sides increasing at different rates.

This characteristic shape is the basis of *optimal path* analysis. Note that the straight line between the two points in the simple proximity bowl in figure 5.1 is the steepest downhill path along the surface, much like water running down the surface. This steepest-downhill path retraces the route of the wavefront that got to the location first. In this case, it is the shortest straight line. Note the similar path indicated on the warped bowl (in the bottom right in fig. 5.1). It goes straight to the barrier's corner, then straight to the starting point, just as you would bend the ruler (if you could). If relative barriers were considered, the path would bend and wiggle in seemingly bizarre ways as it retraced the wavefront (optimal path). This type of routing characterizes the final expansion of the concept of distance—from distance to proximity to movement and finally to *connectivity*, the characterization of how locations are connected in space. Optimal paths are just one way to characterize these connections.

No, business is not as usual with GIS. Our traditional concepts of map analysis are based on manual procedures or their direct reflection in traditional mathematics. Whole procedures and even concepts, such as distance always being the shortest straight line between two points, are coming under scrutiny. Article 6 investigates applications of optimal path analysis, as well as other forms of connectivity.

(Note: See topic 9 for a detailed discussion of the distance algorithm.)

Distance Measurement

(Optimal Path Density and Nth Best Path Analysis)

Keep It Simple, Stupid (KISS).
But it's stupid to keep it simple if Simplifying
Leads to Absurd Proposals (SLAP).

Distance measurement has now been described in new and potentially unsettling ways. Simple distance, as implied by a ruler's straight line, was expanded to effective proximity, which responds to a landscape's pattern of absolute and relative barriers to movement. Under these conditions the shortest line between two points is rarely straight. And even if it is straight, the line's geographic length may not reflect a meaningful measure. How far it is to the airport in terms of time often is more useful in decision making than just mileage. Nonsimple, effective distance is like using a rubber ruler you can bend, squish, and stretch through barriers, like the various types of roads you might use to get to the airport.

The concept of delineating a line between map locations, whether straight or twisted, is termed *connectivity*. In the case of effective distance, it identifies the optimal path for moving from one location to another. To understand how this works, you need to visualize an accumulation surface—described in excruciating detail in the previous article as a bowl-like surface with one of the locations at the bottom and all other locations along rings of successively greater distances. It's like the tiers of seats in a football stadium, but warped and contorted due to the influence of the barriers.

Optimal Path Analysis

Also recall that the steepest downhill path along a surface traces the shortest (i.e., optimal) line to the bottom. It's like a raindrop running down a roof—the shape of the roof dictates the optimal path. Instead of a roof, visualize a lumpy, bumpy terrain surface. A single raindrop bends and twists as it flows down the complex surface. At each location along its cascading route, the neighboring elevation values are tested for the smallest value and the drop moves to that location, then the next, and the next, etc. The result is a map of the raindrop's route. Now, conceptually replace the terrain surface with an accumulation surface indicating weighted distance to everywhere from a starting location. Place your raindrop somewhere on that surface and have it flow downhill as fast as possible to the bottom. The result is the shortest, but not necessarily straightest, line between the two starting points. It retraces the path of the distance wave that got there first —the shortest route whether measured in feet, minutes, or dollars, depending on the relative barrier's calibration.

So much for review; let's expand on the concept of connectivity. Suppose, instead of a single raindrop, there was a downpour. Drops are landing everywhere, each selecting their optimal path down the surface. If you keep track of the number of drops passing through each location, you have an optimal path density surface. For water along a terrain surface, it identifies the number of uphill contributors, termed *channeling*. You shouldn't unroll your sleeping bag where a lot of water is channeling or you might be washed to sea by morning. Another interpretation is that the soil erosion potential is highest at these locations, particularly if a highly erodible soil is present. Similarly, channeling on an accumulation surface identifies locations of common best paths, for example, trunk lines in haul road design or landings in timber harvesting.

Weighted Optimal Path Density

Wouldn't you want to site your activity where it is optimally connected to the most places you want to go?

Maybe, maybe not. How about a weighted optimal path density surface? You're kidding, aren't you? Suppose not all of the places you want to go are equally attractive. Some forest parcels are worth a lot more money than others. (If you have seen one tree, you haven't necessarily seen them all.) If this is the case, have the computer sum the relative weights of the optimal paths through each location, instead of just counting them. The result will bias siting your activity toward those parcels you define as more attractive.

Let's make one further expansion, keeping in mind that GIS is beyond mapping as usual (it's spatial data analysis). As previously noted, the optimal path is computed by developing an accumulation surface, then tracing the steepest downhill route. But what about the next best path? And the next? Or the *n*th best path? This requires us to conceptualize two accumulation surfaces, each emanating from one of the end points of a proposed path. If there are no barriers to movement, the surfaces form two perfect bowls of constantly increasing distance.

Interesting results occur if we subtract these surfaces. Locations that are equidistant from both (i.e., perpendicular bisector) are identified as zero. The sign of nonzero values on this difference map indicates which point is closest; the magnitude of the difference indicates how much closer, relative advantage. If our surfaces were more interesting, say travel time from two sawmills or shopping malls, the difference map shows which mill or mall has a travel advantage, and how much of an advantage, for every location in the study area. This technique often is referred to as catchment area analysis and is useful in planning under competitive situations, whether timber bidding or retail advertising.

But what would happen if we added the two accumulation surfaces? The sum identifies the total length of the best path passing through each location. The optimal path is identified as the series of locations assigned the same smallest value, the line of shortest length. Locations with the next larger value belong to the path that is slightly less optimal. The largest value indicates locations along the worst path. If you want to identify the best path through any location, ask the computer to move downhill from that point, first over one surface, then the other. Thus, the total accumulation surface allows you to calculate the opportunity cost of forcing the route through any location by subtracting the length of the optimal path from the length of the path through that location. "If we force the new highway through my property, it will cost a lot more, but what the heck; I'll be rich." If you subtract the optimal path value (a constant) from the total accumulation surface, you will create a map of opportunity cost, the *n*th best path map.

Got it? In the next article we talk about some further extensions of distance measurement, including the concepts of narrowness and viewshed.

(Note: See topic 9 for a detailed discussion of the optimal path algorithm.)

7 There's a Problem Having All This Sophisticated Equipment

(Establishing Narrowness and Visual Connectivity)

❖   ❖   ❖

The previous article on distance measurement established that distance is simple when we think of it solely in the context of a ruler and shortest straight line between two points. The realistic expansion of distance to consider barriers of movement brought on a barrage of new concepts—accumulation surface, optimal path, optimal path density, weighted optimal path density, *n*th best path. Whew!

Narrowness

Let's get back to some simple and familiar concepts of connectivity. For example, take *narrowness*, or—the shortest cord through a location connecting opposing edges. As with all distance-related operations, the computer first generates a series of concentric rings of increasing distance about a point. This information is used to assign distance to all of the edge locations. Then the computer moves around the edge totaling the distances for opposing edges until it determines the minimum, the shortest cord. For a boxer, the corners of the boxing ring are the narrowest. A map of the boxing ring's narrowness would have values at every location indicating how far it is to the ropes. Small values identify areas where you might get trapped and ruthlessly bludgeoned.

But consider Bambi and Momma Bam's perception of the narrowness of an irregularly shaped meadow. The forage is exceptional, sort of the Cordon Bleu of deer fodder. Its acreage times the biomass per acre suggests that a herd of 50 could be supported. However, the spatial arrangement of these acres may be important. Most of the meadow has large narrowness values—a long way to the protection of the surrounding forest cover. The timid herd will forage along the edges, so at the first sign of danger they can quickly hide in the woods. Only pangs of hunger drive them to the wide, open spaces where Bambi may be lost to wolves—not what you had in mind.

Connectivity

Now raise your sights from cords to rays in three-dimensional space—line-of-sight connectivity or viewshed analysis. Again, concentric rings form the basis of the distance-related algorithm. In this case, as the rings radiate from a starting point (viewer location) they carry the tangent (angle of line between the viewer and a location) that must be beat to mark a location as seen. Several terrain and viewer factors affect these calculations. Foremost is a surface map of elevation. The starting point and its eight surrounding neighbors' elevations establish the initial ring's tangents (rise-to-run ratio, computed as the difference in elevation divided by the horizontal distance). The next ring's elevations and the distance-to-viewer are used to calculate their tangents. The computer then tests if a location's computed tangent is greater than the previous tangent between it and the viewer. If it is, it's marked as seen and the new tangent becomes the one to beat. If not, it's marked as not seen and the previous tangent is still the one to beat.

However, elevation alone is rarely a good estimate of actual visual barriers. Screens, such as a dense forest canopy, should be added to the elevation surface. Viewer height, such as a 90-foot fire tower, also should adjust the elevation surface. Similarly, there may be features, such as a smokestack and plume, that rise above the surface but don't block visual connectivity behind them. When testing if the feature can be seen, this added height is considered, but the enlarged tangent is not used to effectively block locations beyond it. Picky, picky, picky. Yet, to not address the real complexity is unacceptably simplistic. Even more important is to expand the concept of visual connectivity from a point to a set of points forming extended viewers. What is the viewshed of a road, or a set of houses, or powerline, or clearcut? In this case, the extended feature is composed of numerous viewing elements (like the multiple lenses of a fly's eye), each marking what it can see. The total area seen is the collective viewshed.

Ready for another conceptual jump? How about a visual exposure density surface? In this instance, don't just mark locations as seen or not seen but count the number of times each location is seen. "Boy, it would be political suicide to clearcut this area, it's in the view of more than 100 houses. Let's cut over here, only a few houses will be affected." Or, consider a weighted visual exposure surface. This involves marking each location seen with the relative importance weight of the viewer. "Of this area's major scenic features, Pristine Lake is the most beautiful (say 10), Eagle Rock is next (say 6), Deer Meadow is next (say 3), and the others are typical (say 1)." In this case 10, 6, 3, and 1 are added to every location visually connected to the respective features. How about a net-weighted visual exposure density surface? "Joe's junk yard is about the ugliest view in the area (say 10)." If a location is connected to Pristine (ah!), but also connected to Joe's (ugh!), its net importance is zero. It's not as good a place for a hiking trail as just over the ridge that blocks Joe's view, but it's still in sight of Pristine Lake.

This topic has addressed distance and connectivity capabilities of GIS technology. Be honest, some of the discussion is a bit unfamiliar in the context of your current map processing procedures. Yet, I suspect this uncomfortable feeling is more from "I have never done that with maps" than "You can't or shouldn't do that with maps." We have historically developed an ingrained map analysis methodology that reflects the analog map. In doing so, we had to make numerous simplifying assumptions, such as all movement is as straight as a ruler. But GIS maps are digital, and we need to reassess what we can do with maps. Topic 3 contains a series of articles that focuses on a different class of analytic operations that characterize cartographic neighborhoods.

(Note: See topic 9 for a detailed discussion of the visual connectivity algorithm.)

Recommended Reading

Books

Aronoff, S. "GIS Analysis Functions." Chapt. 7 in *Geographic Information Systems: A Management Perspective*. Ottawa, Canada: WDL Publications, 1989.

Gatrell, A.C. "Concepts of Space and Geographical Data." In *Geographical Information Systems: Principles and Applications*, ed. D.J. Maguire, M.F. Goodchild, and D.W. Rhind, Vol. 1, 199-34. Essex, UK: Longman, 1991.

————. *Distance and Space: A Geographical Perspective*. Oxford, UK: Oxford University Press, 1983.

Star, J., and J. Estes. "Manipulation and Analysis." Chapt. 8 in *Geographic Information Systems: An Introduction*. Englewood Cliffs, NJ: Prentice Hall, 1990.

Tomlin, C.D. "Characterizing Locations within Neighborhoods." Chapt. 5 in *Geographic Information Systems and Cartographic Modeling*. Englewood Cliffs, NJ: Prentice Hall, 1990.

Journal Articles

Berry, J.K. "A Mathematical Structure for Analyzing Maps." *Journal of Environmental Management* 11(3): 317-25 (1987).

Eldridge, J., and J. Jones. "Warped Space: A Geography of Distance Decay." *Professional Geographer* 43(4): 500-511 (1991).

Muller, J.C. "Non-Euclidean Geographic Spaces: Mapping Functional Distances." *Geographical Analysis* 14: 189-203 (1982).

Senior, M.L. "From Gravity Modeling to Entropy Maximizing: A Pedagogic Guide." *Progress in Human Geography* 3: 179-210 (1979).

TOPIC 3

ROVING WINDOWS: ASSESSMENT OF NEIGHBORHOOD CHARACTERISTICS

> Imagination is more important than information.
> —*Einstein*

The information surrounding a point often provides insight into spatial problem solving. Neighborhood summaries can be derived from surface configuration to produce slope, aspect, and profile maps. Or, the summaries can relate to the context of the neighborhood for such procedures as spatial interpolation, smoothing, and diversity analysis. More than any other class of operations, roving windows provide entirely new tools and applications for map analysis.

8

Imagination vs. Information

(Characterizing Surface Configuration: Slope, Aspect, and Profile)

❖ ❖ ❖

Einstein is right about imagination, but directed
imagination needs the best information it can get.
–*J.K. Berry*

When viewing a map, the human mind nearly explodes with ideas
about the landscape. Although the ideas are limitless, our ability to
process the detailed spatial data is limited. When the computer
views a map, it sees an organized set of spatial data ripe for processing but has no idea of their significance. Think about it; when
was the last time you took your computer for a walk in the woods?

That's the beauty of the man/machine bonding in GIS. The
user's imagination is magnified manyfold by the machine's ability
to assemble detailed information as directed. Ian McHarg vividly
makes this point in his lectures on GIS when he says, "It is a tool
that extends the mind." We easily conceptualize scenarios for a
landscape but lack the facility to effectively evaluate their relative
merits. That's why we need our little silicon subordinate to take
care of the details. From this perspective, GIS is less computer mapping and spatial database management than it is a decision support
system (DSS) for modeling and evaluating alternative land uses.

Spatially Defined Neighborhoods

The foundation of this thinking with maps is rooted in the analytic capabilities of GIS. Topic 2 describes how our simple concept of
distance has been extended by the computer's ability to calculate
proximity, movement, and connectivity. Now we will investigate a
related set of analytic tools concerned with vicinity or, more technically stated, the analysis of spatially defined neighborhoods.
These procedures consider a map location within the context of its
neighboring locations. As with all GIS processing, new values are

computed as a function of the values on another map. Two steps are involved in neighborhood analysis. First, establish the neighborhood and its values; then, summarize the values.

Determination of neighborhood membership is like a roving window moving about a map. Picture a window with nine window panes looking straight down onto a piece of the landscape. (Sort of makes you feel all powerful, doesn't it?) Now, like a nosey neighbor, systematically move it around to check out the action. Suppose your concern is surface configuration. You would note the nine elevation values within the window, then summarize the three-dimensional surface they form. If all the values were the same (e.g., say 100-foot elevation), you would say it was a boring, flat area and move your window slightly to one side. Some larger values appear on the side window panes. Move it another couple of notches and the window is full of different elevation values.

Generating a Slope Map

Imagine that the nine values become balls floating at their respective elevations. Drape a sheet over them like the magician places a sheet over his suspended assistant. (Who says GIS isn't at least part magic?) There it is, surface configuration. Now numerically summarize the lumps and bumps formed by the ghostly sheet. That means reducing the nine values to a single value characterizing the surface. How about its general steepness? You could compute the eight individual slopes formed by the center value and its eight neighbors (change in elevation divided by the change in horizontal distance expressed as a percent). Then average them for an average slope. You could, but how about choosing the maximum slope? That's what water does. Or the minimum slope? That's what a weary hiker does. In special cases, you would choose one of these statistics.

Most often you are interested in the best overall slope value. This is determined by the best-fitted plane to the data. Replay your vision of the nine floating balls. Now insert a glass panel (plane) in such a way that the balls appear balanced about it, minimizing the deviations from the plane to the balls. If you're a techy, you will recognize this as a simple linear regression, except in three-dimensional space. But, if you value your computer's friendship, don't use a least-squares fit algorithm. Use vector algebra; it's much faster.

As the window progresses about the map, slope values are assigned to the center window pane (cell) until all locations have received a value—abracadabra, a slope map. Locations with larger values indicate steep terrain, smaller values for gently sloped terrain. But what is the terrain's orientation? That's an aspect map. Move the same window and best-fitted plane about the map, but this time use its direction cosines to indicate the orientation of the plane. Is it facing south? Or north? Or 47º azimuth? It could make a big difference. If you're trying to grow trees in a moisture-limited region, those south-facing slopes are only good for rattlesnakes. However, if there is ample water, they get the most sunlight and tend to grow the best trees. If you're a land planner, the southern slopes tend to grow the best houses, or at least, the lowest heating bills.

Generating a Profile Map

One final surface configuration factor to consider is profile. Imagine a loaf of bread, fresh from the oven. It's like an elevation surface. At least mine has deep depressions and high ridges. Now start slicing the loaf and pull away an individual slice. Look at it in profile, concentrating on the line the top crust portion traces. From left to right, the line goes up and down in accordance with the valleys and ridges it sliced through. Use your arms to mimic the shapes along the line. A V with both arms up for a valley. An inverted V with both arms down for a ridge. Actually there are only nine fundamental profile classes (distinct positions for your two arms). Values one through nine will serve as our numerical summary of profile.

However, a new window is needed. This time, as you look down onto the landscape, move a window with just three panes along a series of parallel lines. At an instant in time, you have defined three elevation values. Compare the left side value to the center. Is it higher or lower? Put your left arm in that position. Now do the same for the right side and center values. Note the fundamental profile shape you have formed and assign its value to the center location. Move the window over one pane and repeat until you have assigned a profile value to every map location. The result of all this arm waving is a profile map, the continuous distribution profiles viewed from some direction. Provided your elevation data are at the proper resolution, it's a big help in finding ridges and valleys running in a certain direction.

That's where the gold might be. Or, if you look from two opposing directions (orthogonal) and put the two profile maps together, a location with an inverted *V* in both directions is likely a peak.

There is a lot more to neighborhood analysis than just characterizing the lumps and bumps of the terrain. What would happen if you created a slope map of a slope map? Or a slope map of a barometric pressure map? Or of a cost surface? What would happen if the window wasn't a fixed geometric shape, say a 10-minute drive window? I wonder what the average age and income is for the population within such a bizarre window?

(Note: See topic 9 for a detailed discussion of the slope-aspect algorithm.)

9

It's Like New Math; I'm Just Too Old

(The Map Derivative and Its Use)

❖　　❖　　❖

The previous discussions have established maps as data. That maps are numbers in a GIS is what allows us to go beyond mapping. It extends pens, symbols, and colors to spatial statistics, mathematics, and modeling. But in doing so, does it leave the typical user in the dust?

Your initial response may be, "You bet. It's like new math; I'm just too old." We have established procedures for dealing with maps and data, built up through years of study at the school of hard knocks. Maps are colorful graphics you hang on the wall, and data are colorless numbers you align in columns. The thought that maps and data are one and the same is unsettling. Well, let's return to the Land of Oz, featuring neighborhood characterization.

Recall the discussion in the previous article; the procedure for assessing surface configuration was described as a window with nine panes moving around on a map of elevation values. At any one instant in time, the nine values in the window are summarized for the slope and aspect of the three-dimensional surface they form (see article 8 for algorithms). The results are assigned to the location at the center of the window and the window then advances to the next position. This procedure is repeated until a slope or aspect value is assigned to all the locations in a project area. Useful information, but is that all there is?

Of course not, this is the Land of Oz. A slope map is actually the first derivative of the elevation surface. You remember the derivative; that blood-sucking calculus teacher threatened you with it. In simple terms, a derivative indicates the rate of change in one variable with respect to another. In the terrain slope example, it is the rate of change in elevation per geographic step—rise is to run. If elevation doesn't change, the terrain is flat. If elevation changes a lot over a short distance, the terrain is steep. Slope (derivative) indicates how rapidly surface features are changing throughout your map. Aspect indicates the direction of this change. "I can handle that," you say.

Surface Roughness Map

OK. Then, what is the second derivative of an elevation map? The slope of a slope map? The first derivative is the rate of change in elevation per geographic step. Then, the second derivative must be the rate of change in the rate of change in elevation per geographic step. What? That doesn't make sense. Maps are maps and math is math and you shouldn't confuse the two. The second result is called a *surface roughness map*. It shows you where those little brown contour lines are close together, then far apart, then close together again, then far apart —heartbreak terrain for tired hikers. Close your eyes and envision a steep mountainside you have to climb. It could be worse.

Suppose the slope isn't constant (a titled, straight line in profile), but variable (a titled, wiggly line in profile). To get to the same point, you would hike up, then down, then up, then down. At each rise, your hopes (and elevation advantage earned) would be lost. Neither man nor machine likes to run around in this sort of terrain. If you're a forester, a roughness map could help you in estimating harvesting costs. If you're a regional planner, it could help you assess possible corridors for a proposed highway. If you're a hydrologist, it could help in modeling surface runoff.

Weather predictors have used map derivatives for years. They collect barometric pressure readings at weather stations and then interpolate these data into pressure gradient maps. To the casual observer, these maps look just like a terrain surface connected by peaks, valleys, and a host of varying slopes. Winds blow from high to low pressure, from the peaks to the valleys along the pressure gradient. The steeper the slope between peaks and valleys, the faster the wind. Therefore, a slope map of the pressure gradient surface indicates wind speed at each location. The aspect of the same map indicates wind direction at each location. So that's how weather predictors get tomorrow's gusty prediction. Or how about a cooling pond's thermal gradient, a mountain of high temperature at the point of discharge that dissipates at different rates and directions as a function of depth, bottom conditions, streams, and springs. Slope and aspect of this thermal gradient map these complex interactions.

Let's recapitulate before we go on. The familiar concepts of terrain slope and aspect are down-to-earth examples of that elusive mathematical concept of the derivative. It gives a firm footing in the real world to one of the most powerful mathematical tools in numerical space. A slope map indicates the spatial distribution of the rate of change in any map variable. An aspect map indicates the direction of that change. Because slope and aspect have a more general meaning than giving direction to raindrops, they can be used for a number of other purposes.

Travel-Time Map

Consider an accumulation surface. Remember that bizarre map discussed in topic 2 in connection with weighted distance measurement? For example, a travel-time map indicates how long it would take to travel from one location to all other locations in a project area. If you move in straight lines in all directions, a perfect bowl of constantly increasing distance is formed. However, if you are a realistic hiker, your movement throughout the area will bend and twist around both absolute and relative barriers, as defined by terrain and land cover features. The result will be a travel-time map that is bowl-like, but wrinkled with ridges and valleys. Weird, but still a map surface that is not unlike a terrain surface. Its slope indicates the ease of optimal movement across any location in the project area and its aspect indicates the direction of optimal movement. So what?

Think about it. This is a map of the pattern of optimal movement in an area. Such information is invaluable whether you are launching a crew of fire fighters or a cruise missile. But why stop at a travel-time map? Why not create a cost surface in which the derivative produces a marginal cost map – the cost to go an additional step in space? Or a marginal revenue map? Such is the decision fodder that fuels the salaries of most MBAs, only expressed as an entire map instead of a single number.

At least two things should be apparent from the preceding discussion. First, map analysis in a GIS is based on mathematics. Maps are large groups of numbers and most of the traditional analysis techniques are applicable. Certainly, the derivative is a switch hitter that propels GIS beyond mapping. How about the integral? Sure, why not? Envision a single, huge window that covers the entire map and sums all the values. How about the mean? Just divide the integrated sum by the number of locations. How about the standard deviation? And the coefficient of variation? Sure, but that's for the next article. The other thing that should be apparent is that you may not want to have a thing to do with this "map-ematics." Maybe, maybe not. If you've read this far, a seed has been planted. At minimum, the thought of a map derivative will haunt you in the shower. Like that innocent bather who checked in at the Bates Motel.

10 Torture Numbers; They'll Tell You Anything

(Basic Considerations in Spatial Interpolation)

❖   ❖   ❖

You cannot cross a chasm in two jumps.
—*Old Russian Maxim*

The previous two articles introduced the idea of roving a small window throughout a map, summarizing the surface configuration detected at each location. Slope, aspect, and profile of an elevation surface made sense. You dug in your fingernails on steep, southerly slopes and pitched your tent on flat ones. But the extension of this concept to abstract maps, such as travel time and cost surfaces, may have been a bit uncomfortable. Relating it to math's derivative made it downright inhospitable. Hopefully, you saw through all that academic hyperbole and visualized its application potential. Surface configuration tells you how and where a mapped variable is changing, important information to tell you how and where you should change your management action.

There is another fundamental way to summarize a roving window—statistically. For example, you might ask, "How many houses are there within a quarter-mile radius?" or, "What is the average lead concentration in the soil within a hundred meters?" In these instances, the numbers defining a map are windowed, then statistically summarized, with the summary value assigned to the center location of the window. The window repeats this process as it moves about the map. The list of data summary techniques is large indeed —total, average, standard deviation, coefficient of variation, maximum, minimum, median, mode, diversity, deviation, and many others. In fact, most traditional mathematical and statistical operations can be used to summarize the data in the window.

Weighted Windows

But that's not all. Because of the spatial nature of mapped data, new operations arise, such as Fourier two-dimensional digital filtering, a real trek on the quantitative side that is beyond the scope of this book. Yet the basic concept embedded in these seemingly complex procedures actually is quite simple. Consider the housing density map noted previously. The number of houses within a quarter mile of any location is an indicator of human activity. More houses mean more activity. Yet suppose your concern is a noisy neighborhood. It's not just the total number of houses in the vicinity of a location, but their juxtapositioning. If the woofers and tweeters are concentrated close to you, you'll be rockin' through the night. If most are at the edge of your neighborhood window, no problem. Physics describes this condition as the dissipation of sound, as a nonlinear function of distance. You probably describe it as relief. That means a house twice as far away sounds a lot quieter than the one next door.

To our GIS, that means a distance-weighted window capability. Weights are calculated as an inverse function of the distance for each window position. The result is a matrix of numbers analogous to writing a weighting factor on each pane of glass forming the entire window. When you look through this window at the landscape, multiply the data you see by their respective weight, then statistically summarize these data and assign the summary value to the center location of the window. In this instance, the noise emanating from each house is adjusted for its positioning in the window and the total noise computed by summing all the adjusted values.

Spatial Interpolation

The concept of weighted windows is easy to grasp. The procedures used to derive the weights is what separates the manager from the mathematician. For now, let's stick to the easy stuff. For example, consider weighted-nearest-neighbor interpolation. It uses the same inverse distance-squared window as described above. Instead of noisy data, field-collected measurements of well pollution levels, barometric pressure, or animal activity can be used. Consider figure 10.1. Inset (a) shows a geographic plot of animal activity recorded during a 24-hour period at 16 sample sites for a 625-hectare project area. Note the higher measurements are concentrated in the northeast while the lower measurements are in the northwest. The highest activity level is 87 while the lowest is 0, forming a rather large data range of 88. The computed average activity is 22.56, with a standard deviation of ± 26.2. On the whole, the area is fairly active but not too bad.

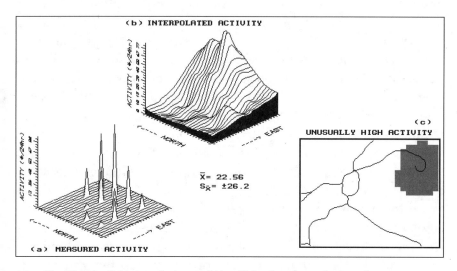

Fig. 10.1. Spatial interpolation and identifying locations of unusual response.

Inset (b) shows the result of moving the inverse distance-squared window over the data map. At each stop the activity data are multiplied by their weighting factor, and the weighted average of all the adjusted measurements is assigned to the center of the window. This provides an estimate (interpolated value) of activity that is primarily influenced by those sampled points closer to it. It's common sense; if there is a lot of activity immediately around a location, chances are there is a lot of activity at that location. This often is the case, but not always (that's why we need the mathematician's complex weighting schemes).

Whoa! You mean tabular data can be translated into maps? In many instances, yes. The 22.56 average animal activity actually implies a map. It's just that it is a perfectly flat surface, estimating that 22.56 activity level is everywhere, plus or minus the standard deviation of 26.2, of course. But it doesn't indicate where you would expect more activity (plus), or where to expect less (minus). That's what the interpolated map does. If higher activity is measured all around a location, such as in the northeastern portion, then why not estimate more than the average? Not a bad assumption in this case, but it depends on the data. As with all map analysis operations, you aren't just coloring maps, you're processing numbers with all the rights, privileges, and responsibilities of mathematics and statistics. Be careful.

You might be asking yourself, "If the interpolated surface predicts a different animal activity at each location, I wonder where there are areas of unusual activity." That's a standard normal variable (SNV) map. It's this simple: SNV = [(x-average)/standard deviation]*100, where x is an interpolated value. It's not as bad as you might think. If the interpolated value (x) is exactly the same as average, then it computes to zero—exactly what you would expect. Positive SNV values indicate areas above the average (more than you would expect); negative values indicate areas below the average (less than you would expect). A +100 or larger value indicates areas that are 100 percent or more of a standard deviation above the average—unusually high activity. Figure 10.1c locates this area as easily accessible by the woods road in the northeast. Now get in your pickup truck and check it out. For the techy types, the SNV map is the geographic plot of the standard normal curve and the map in inset (c) is the plot of the upper tail of the curve. For the rest of us, it's just a darn useful technique that provides a new way of looking at our old data. It brings statistics down to earth.

So, spatial interpolation is a neighborhood operation involving, at least conceptually, a roving window, and a weighted one at that. Actually, it's an operation fairly similar to the familiar concepts of slope and aspect calculation. In the next article, we finish our brush with neighbors by considering dynamic windows. Once you have tasted weighted windows, you will love dynamic ones.

11 I Don't Do Windows

(Summarizing Weighted Windows)

The previous two articles described neighborhood operations as moving a window around a map. We found that the data within a window at an instant in time could be used to characterize the surface configuration (e.g., slope or aspect) or generate a summary statistic (e.g., total or average). The value representing the entire neighborhood is assigned to the window's focus, then the window shifts to the next location.

This is a simple, straightforward process, except on two counts. One is understanding the wealth of mathematical and statistical processes involved. Most traditional math/stat operations are possible (those termed *commutative* operations for the techy types). That leads to the other complicating count—why would I want to do these unnatural, numerical things to a map? And what would I do with the bizarre results, such as a marginal cost map (slope of a cost surface)? Hopefully, articles 9 and 10 provide enough examples to stimulate your thinking beyond traditional mapping to maps as data and, finally, to map analysis.

Intellectual stimulation, however, can quickly turn to conceptual overload. Risking this, let's return to spatial interpolation. Recall that interpolation involves moving a window about a map, identifying the sampled values within the window, summarizing these samples and finally assigning the summary to the window's focus. The summary could be a simple arithmetic average or a weighted average, most commonly the inverse distance-squared weighted average.

Directional Window Weights

How about another conceptual step? Instead of making the weights a simple function of distance, incorporate a bias based on the trend in the sampled data. This is what that mysterious interpolator kriging does. It's based on common sense—the accuracy of an estimated value is best at a sampled location and becomes less reliable as interpolated points get farther away. Simple and straightforward. But the direction to a sampled value often makes a difference. For example, consider the change in ecological conditions as you climb from Death Valley to the top of Mount Whitney. As elevation rapidly increases, you quickly pass through several ecological communities. If you move along an elevation contour, things don't change as quickly. For years ecologists have used elevation in their mapping.

Now envision a map of this area (or refer to a map of southeastern California). Major changes in elevation primarily occur along the east/west axis. Most of the contours (constant elevation) run along the north/south axis. If our understanding of ecology holds, an estimated location should be influenced more by samples in a north/south direction from it. Samples to the east/west should have less influence. That's what kriging does. It first analyzes the sample data set for directional bias, then adjusts the weighting factors it uses in summarizing the samples in the window. In this case, it would uncover the directional bias in the sample data (induced by elevation gradient, provided theory holds), then set the window weighting factors.

Another way to conceptualize the direction-biased window is as an ellipse instead of a circle. Inverse distance-squared weighting forms concentric halos of equal weights—a circular window. Kriging windows form football-shaped halos reaching out the farthest in the direction of trend in the data—an elliptical window. For the techy few, this is similar to the Mahalanobis distance in multivariate analysis. For the rest of us, it demonstrates the first consideration in roving-window design—direction. Why do windows have to form simple geometric shapes, like circles and squares, in which all directions are symmetrically considered? Well, they don't.

For example, consider secondary source air pollution and health risk mapping. If you have a map of the lead concentration in soils, you might identify as risky those areas with high concentrations within 500 meters. To produce this map you could move a window with a radius of 500 meters throughout the lead map, assigning the average concentration as you go. But this process ignores the prevailing winds. An area might have a high concentration to the north, with low concentrations elsewhere. Its average might be within the guidelines, but the wind blows from the north. The real effect would be disastrous for a home built at this location. In this case, a wedge-shaped window oriented to the north (upwind) would be more appropriate.

Effective Distance Weights

Actually, there is more to windows than just direction. There is distance. For example, consider a big wind from the north. Under these conditions, relatively distant locations of high concentrations would affect you. Under light winds, they wouldn't. Considering both wind direction and strength results in a dynamic window that adjusts itself each time it defines a neighborhood. To accomplish this, you need a wind map (often referred to as a wind rose) as well as the lead concentration map. The wind map develops the window configuration, and the lead map provides the data for summary. In reality, cumulative effects and particulate mixing should be considered, but that's another, even more complicated, story. In the end, it simply results in better definition of window weights.

Let's try another dynamic window example. Suppose you were looking for a good place for a fast food restaurant. It should be on an existing road (the automobile is king). It should be close to those most prone to a Mac attack (wealthy families with young children). Armed with these criteria, you begin your analysis. First you need to build a database containing information on roads and demographic information. With any luck, the necessary data are available in the TIGER/Line Files available for your area.

Now all you need is a procedure that relates movement along roads from a location to the demographic data—a travel-time window. Based on the type of roads around a location, move out 10 minutes in all directions. The result is a spiderweb-like window that reaches farther along fast roads than along slow roads. A bit odd-shaped, but it's a window no less. Now lay the window over the demographic data to calculate the average income and number of children per household. Assign your summary value, then move to the next location along the road. When all locations have been considered, the ones with the highest yuppie indexes are candidates for restaurant locations.

All this may sound simple (ha!), but it's a different story when you attempt to implement the theory. Several GIS software packages allow you to create dynamic weighted window maps. It's not a simple keystroke, but a complex command "macro." Such concepts are pushing at the frontier of GIS. It's currently the turf of the researcher. Then again, GIS as you know it was just a glint in the researcher's eye not so long ago. You will probably do windows in your lifetime.

Recommended Reading

Books

Burrough, P.A. "Methods of Spatial Interpolation." Chapt. 8 in *Principles of Geographical Information Systems for Land Resources Assessment*. Oxford, UK: Oxford University Press, 1987.

Cressie, N. *Statistics for Spatial Data*, New York: Wiley, 1991.

Flowerdew, R. "Spatial Data Integration." In *Geographical Information Systems: Principles and Applications*, ed. D.J. Maguire, M.F. Goodchild, and D.W. Rhind, Vol. 1, 375-87. Essex, UK: Longman, 1991.

Rippley, B.D. *Spatial Statistics*. New York: Wiley, 1981.

Tomlin, C.D. "Cartographic Modeling." In *Geographical Information Systems: Principles and Applications*, ed. D.J. Maguire, M.F. Goodchild, and D.W. Rhind, Vol. 1, 361-74. Essex, UK: Longman, 1991.

———. "Characterizing Locations within Neighborhoods." Chapt. 5 in *Geographic Information Systems and Cartographic Modeling*. Englewood Cliffs, NJ: Prentice Hall, 1990.

Journal Articles:

Burgess, T., and R. Webster. "Optimal Interpolation and Isarithmic Mapping of Soil Properties: The Semi-Variogram and Punctual Kriging." *Journal of Soil Science* 31: 315-31 (1980).

Hubert, L., et al. "Generalizing Procedures for Evaluating Spatial Autocorrelation." *Geographical Analysis* 13: 224- 33 (1981).

Foster, S.A. "The Expansion Method: Implications for Geographic Research." *Professional Geographer* 43(2): 131-42 (1991).

McDonnell, M.J. "Box Filtering Techniques." *Computer Graphics and Image Processing* 17: 65-70 (1981).

Webster, R., and T. Burgess. "Optimal Interpolation and Isarithmic Mapping of Soil Properties: Changing Drift and Universal Kriging." *Journal of Soil Science* 31: 505-24 (1980).

Topic 4

What GIS Is and Isn't: Spatial Data Mapping, Management, Modeling, and More

If you can't win on one side of an issue, try both.
—Old Political Axiom

Most initial applications of GIS automate current cartographic practices. However, the greatest return on investment in GIS technology is realized through entirely new applications inspired by the new set of map analysis tools. This section develops an awareness of the considerations and conditions that move user perspective from computer mapping to spatial database management, to application modeling and beyond.

12   Technobabble

(GIS Is More Than the Automation of Current Procedures)

A seemingly endless drone masking what otherwise would be a clear understanding of technology.
–J.K. Berry

You've read about some strange things in the Beyond Mapping series: "map-ematics," effective distance, map derivative, weighted windows, optimal path density, and net weighted visual exposure density surface, just to name a few. "You can't do that to a map; that's disgusting; are you sure it's legal?" may have been some of your comments. Technobabble. Just a bunch of technobabble.

Professor Arthur Hough of San Francisco State University's Communication Arts Department sums it up, "GIS is not just warm, woolen socks." Hough explains that GIS is a change in mapping (and communications, for that matter) like the cocoon's change to the caterpillar and butterfly. Ugly, but effective. To those on the outside, the cocoon appears to just sit there. But on the inside, there is total upheaval and complete restructuring. Such is the metamorphosis brought on by the digital map.

Maybe, maybe not. Most of the practical applications of GIS involve automating current manual procedures. Correct that; most involve investing in a database that, hopefully, will eventually automate the current manual procedures. This is a lot of work, and we've been doing it for years. The perceived benefit, once GIS is on-line, is that we can do it faster, with greater detail and in vibrant colors. The butterfly is obviously superior to the worm. But more importantly, it is radically different. Understanding the differences and developing new procedures is the behind-the-scenes revolution of GIS in the 1990s. It's not business as usual.

Consider the familiar map overlay operation. Suppose you are interested in locating your company's forest management parcels containing both douglas fir trees and Cohasset soil. In the 1960s, your cartographic solution was to overlay both maps on a light-table and sketch. The result was a single map depicting the intersection—good spatial characterization. However, if acreage estimates were required, hours of planimeter or dot grid work were required.

Your statistical procedure more likely involved searching a database. Information such as acreage, timber, and soil types for each management parcel was written on a card. Holes were punched along the edges to summarize the information. A geographic search simply involved passing a long needle through the appropriate edge position. When lifted, the parcels meeting that condition fell out of the stack. Repeat with the "sieved" subset, and the cards containing both conditions fell out, douglas fir and Cohasset soil. Add up the acreage—good statistical characterization.

A couple of problems persisted. The procedures were tedious and disjointed. You spent hours drafting and calculating for even a simple query over a small area. So what? The procedures were as comfortable as a pair of warm, woolen socks. Some folks even argued that the time involved was peanuts in comparison to the hours of creating, caressing, and cursing an automated database. That is a valid argument. Each of us knows our most efficient mode.

But the problem of the two manual systems being disjointed is critical. It is not just a matter of time; it's the nature of the information derived—an answer to a specific question, a dead end. The information doesn't become part of the database. It cannot be easily shared with others and their subsequent analyses incorporated. Your drafting and calculating, for all you know, may be repeated the following week by a colleague down the hall. In large part, the computer's capacity to store and share information is what tipped the scales from index cards to database management. At least in the beginning, it wasn't efficiency and ease of use. The transition was (is?) painful for most of us. Now the Information Age is being heralded as the modern equivalent of the Industrial Age, not merely a progression of technology as much as a radical departure.

It's like the automobile. At first, it was just an engine affixed to a wagon. Aspirations for the newfangled thing were for it to do the work of a team of horses. Nothing more, nothing less. But as the car evolved, new demands for speed and capacity continuously redefined the "automated wagon." Entirely new concepts such as aerodynamics, four-wheel drive, and catalytic converters have become commonplace. Nostalgia aside, isn't the car a vast improvement on the wagon? Though different, it's not all that complicated to learn to drive a car as compared to driving a team of horses. You get there a lot quicker, and you can do more.

The transition from the horse to horsepower, however, required both personal and social investments. GIS is making similar demands. Your challenge is to understand the differences between traditional mapping processes and apply these new procedures in creative ways. But that's not enough. It's like the 80-mph Bugatti—awesome in its time yet useless without paved, 80-mph roads. A washboard wagon trail not only limits your potential; it's downright dangerous.

That's where we are with GIS. The equivalent of supersonic (or supernatural, your choice) procedures for map analysis are in place. Most folks opt to keep things down to earth and apply GIS in traditional ways. Those who choose the high road soon find that the base data are as rocky as a wagon trail. Our historical concepts of mapped data are rooted in the map as a generalized image for human viewing. But GIS considers maps as large sets of numbers poised for quantitative analysis. Creation of an image is just one of the things it can do. Statistical and mathematical analytics comprise a multitude of other things.

From this perspective, GIS is the blend of cartographic and numerical processing that was missing in the 1960s. The concept that it is a "cash register" in which transactions on the landscape are recorded is one manifestation of the link. This definition is tremendously useful, but it challenges neither the basic procedures nor the basic data form. The topics in this book continue to do both. Even something as intuitively obvious as overlaying a couple of maps will be contorted into whether it is a point-by-point, regionwide, or mapwide operation. Concern for both spatial and thematic "error propagation" also is a must. Techno-babble. TECHNOBABBLE! But it's interesting—sort of like those electronic woolen socks with a nine-volt battery you slip into your ski boots.

13 | What's Needed to Go Beyond Mapping?

(GIS Requires Advanced Analytical Capabilities)

❖   ❖   ❖

To boldly go where no one has gone before.
                                        —Captain Pickard

GIS is certainly a workhorse. It manages our spatial data. It provides timely updates to our databases. It creates colorful and valuable map products. In short, it is rapidly becoming an integral part of our record keeping and report generation. However, to some, it is an ill-tempered racehorse, moving at breakneck speed—expensive and cantankerous at best. It is the domain of the overly indulgent rich. To others, it is a Pegasus, whose wings soar us to new heights. It is a radical departure from traditional mapping with entirely new concepts.

In reality, GIS is all of these things. The digital nature of GIS maps provides the skeleton for each perspective. After you have a computer-compatible map, only your imagination limits its use. Well, that and your software vendor. Like horses, GIS software comes in a wide variety of sizes, shapes, and colors. A Clydesdale won't make it at a fox hunt with the queen. Nor is a thoroughbred suited for the plow. In selecting your best-fitted beast, functionality plays a large part in determining its appropriate use. The overlaying and geographic search functions of a GIS are relatively familiar to most current and aspiring users. But what is needed for GIS modeling (map analysis)? What takes a GIS beyond mapping?

A checklist of the analytical capabilities that go beyond the basic GIS procedures follows. It is designed to spur discussion among users and vendors as well as provide a structure for the rest of the articles in this book.

If I Wanted It All.

MATHEMATICAL OPERATORS

- *Basic Math.* The most frequently used buttons on your pocket calculator—add, subtract, multiply, divide, average, etc.

- *Advanced Math.* The rest of the buttons—such as the trigonometric functions, powers, roots, etc.

- *Macro Command Language.* The ability to branch, loop, and test within a sequence of map processing commands.

SPATIAL STATISTICS

- *Descriptive Statistics.* Describe a single map variable or set of map features. For example, a standard normal variable (SNV) map identifies statistically unusual areas and is computed by $SNV = (Xobs-Xmean)/Xstdev)*100$, where *Xobs* is the observed value at a particular location, *Xmean* is the mapwide average of the variable, and *Xstdev* is the standard deviation of the variable. Another example is the calculation of the average for one mapped variable (e.g., slope) for a set of map features (e.g., timber harvest parcels). A basic set of statistical procedures includes the frequency counts, average, standard deviation, coefficient of variation, minimum, maximum, mode, median, diversity, deviation, and proportion similar.

- *Comparative Statistics.* Compare two or more map variables or sets of map features. For example, a simple t-test can be used to compare the similarity (coincidence) between two maps. A basic set of statistical procedures includes simple and frequency-weight crosstabs, chi-square, Scheffe, t-, and F-tests.

- *Predictive Statistics.* Establish relationships among map variables. For example, a simple linear regression can be developed for predicting one map variable from a set of other map variables. A basic set of statistical operators includes clustering and simple linear, multiple, and curvilinear regressions.

DISTANCE MEASUREMENT

- *Simple Distance.* Calculates the shortest straight line between two points (Pythagorean Theorem).

- *Buffer.* Identifies all locations within a specified distance of a point, line, or areal feature, such as all locations within 100 feet of a stream.

- *Narrowness.* Determines *constrictions* as the shortest cord connecting opposing edges of an areal feature, such as forest opening.

- *Simple Proximity.* Identifies the shortest straight-line distance from a point, line, or areal feature to all other locations in a mapped area. This is similar to a series of buffers of equal steps emanating from a set of features (e.g., ripples in a pond).

- *Effective Proximity (Movement).* Identifies the shortest, but not necessarily straight-line, distance from a point, line or areal feature to all other locations in a mapped area. Distance is measured as a function of absolute and relative barriers affecting movement (friction). This is similar to a travel-time map in network analysis, yet movement is allowed over a continuous surface.

NEIGHBORHOOD CHARACTERIZATION

- *Surface Configuration.* Characterizes a continuous surface's form. A basic set includes slope, aspect, and profile. An advanced set includes an array of engineering techniques such as grade, curvature, and cut/fill.

- *Roving Window Summary.* Summarizes values within a specified vicinity around each location, such as the total number of houses within an eighth of a mile creating a housing density surface. A basic set of summary operators includes total, average, standard deviation, coefficient of variation, maximum, minimum, mode, median, and diversity. Statistics comparing the center of window to its neighbors include deviation and proportion similar.

- *Interpolation.* Computes an expected value for each map location (continuous surface) based on a set of point samples. A basic set includes weighted-nearest-neighbor and kriging techniques.

VISUAL EXPOSURE

- *Viewshed Delineation.* Identifies all locations that are visually connected to a point, line, or areal feature.

- *Exposure Density.* Determines how often each location is visually connected to a line or areal feature.

- *Weighted Exposure Density.* Calculates a weighted-visual-exposure value for each map location by considering the relative visual importance of each viewing point, line, or areal feature. For example, an area visually connected to a major highway has a higher exposure value than another area connected the same number of times to a lightly traveled road.

OPTIMAL PATHS

- *Simple Paths.* Determines the best route from one location to another along a set of lines (network analysis) or over a continuous surface. In the case of a terrain surface, it identifies water flow. In the case of a travel-time surface, it identifies the quickest path.

- *Path Density.* Counts the number of paths passing through each element of a network or continuous surface that optimally connect a set's starting and finishing points. In the case of a terrain surface, it identifies confluence (channeling) of water flows. In the case of a travel-time surface, it identifies traffic confluence (heavily used corridors), be it cars, hikers, or elk.

- *Weighted Path Density.* Similar to normal path density, except each path is weighted by the volume or flow along the path.

SHAPE CHARACTERIZATION

- *Convexity Index.* A measure of the boundary regularity of an areal feature based on the ratio of its perimeter to its area.

- *Fractal Geometry.* Quantifies the complexity of a feature's shape as an exponential relationship of its perimeter, area, and fractal dimension.

- *Spatial Integrity.* Measures of the intactness of areal features relating holes and fragments of features forming the map mosaic, such as clear cuts in a forested landscape.

- *Contiguity.* Assesses the pattern among groups of features, such as whether the clear cuts are clumped together or evenly distributed in the landscape.

- *Interfeature Distance.* Computes the average distance within a set of map features, such as individual parcels of endangered species habitat.

Whew! That's all there is to it. At first glance, the advanced analysis grab-bag may seem a bit overwhelming. It appears less of a work-horse than a whole pack team, all pulling in different directions. But that's because it's unfamiliar; not that it's that tough. The majority of the concepts have been in your visceral (if not your conscious) state for a long time. Heaven knows they have been in textbooks for years. They're waiting for you to apply them in innovative ways that take you beyond mapping. Just ask your vendor.

14 Who Says You Can't Teach an Old Dog New Tricks?

(GIS Models Are Flexible, Succinct, and Dynamic)

❖　　❖　　❖

Now that I am older, with several years under (and over) my belt, I am enamored with new tricks, the GIS bag of tricks. I am an old forester who has evolved from a "cyberphobic" to a "cyberphiliac." Nothing is more fun than wrestling with a new perspective on the venerable field of forestry.

Let's talk about the real world, the deep woods, a place of dirt, slopes, trees, and fish. How about a new perspective on the old problem of timber harvesting and fish romance? You know, create a map that says, "If you run your skidder here, it will likely kill the spawning woopie there." (A *skidder* is a tractor-like thing that drags trees around in the woods.)

This harvest-stopping loggerhead between the fish and the forester seems simple—just leave a 100-foot buffer around class two streams. That should save the sexy salmon. Or will it? My bet is about the time you get your harvest plan drafted and hung on the wall, the local angling association will ask, "What would happen if, just to be on the safe side, you considered a 200-foot buffer?" A valid question. A real concern. Worse yet, if you don't address it, chances are you will be asked the same question by a judge at an injunction hearing. But you already wore out a box of crayons (let alone your patience) drafting the 100-foot buffer plan. You don't have time to respond to every little concern.

But, like those dark moments in an old western before the calvary arrives, all is not lost. GIS will save you. Just edit the 100 to 200 in your GIS model and rerun it. It will take you and your silicon subordinate mere minutes to respond. Then you can prove there won't be much change in the plan. Why wouldn't the fishermen and judge believe you? You know these things. You're an old hand when it comes to knowing what animals do in the woods.

GIS is just another hoop to jump through. It's a waste of time. It's a waste of money. But it does cover your . . . ah . . . decisions.

You're a crafty old fox who knows what GIS is and isn't. Old woods wisdom puts this newfangled technology in its place. It's not a new trick, just an accelerated old trick. The old dog is happy with that. But more importantly, are the fish happy? Is your company's bottom line happy? My guess is that your old perspective on map analysis isn't helping. I bet you're killing fish and losing revenue at the same time; you're just doing it faster with your concept of a GIS.

We're not just talking trees and fish here; there is dirt in between. What you need is a realistic sediment loading potential (SLP) model. Let's apply some common sense. The farther away from the stream you keep the skidder, the less the sediment should complicate salmon romance. That's why the fishermen wanted to increase the buffer. But are all buffer feet the same? Not by a long shot. If there are steep slopes of sparsely covered vegetation between your skidder and the fish, you had better be a lot farther away than 200 feet. But if gentle, vegetated terrain separates them, you could harvest well within 100 feet of the stream. Your old trick, whether tediously or rapidly applied, killed the fish and robbed the logger.

You need a new trick, like weighted distance measurement. This topic was driven into the ground in topic 2, but it's definitely applicable here. You need a "rubber ruler" that stretches in erodible places and shrinks in stable ones (see topic 2, article 6 [concept] and topic 9, article 29 [algorithm] for the discussion of rubber rulers). Figure 14.1 shows this effect. The top map calculates sediment loading potential as the inverse of the distance squared. You can do this with a ruler. But you would be hard pressed to create the more realistic bottom map, which considers both intervening slope and cover conditions in assessing effective sediment loading distance.

GIS is more than a faster mapper. It's more than a replacement for your old oak file cabinets. It's a technology providing new tools for resource management planning, such as spatial statistics, effective distance, optimal paths, and visual exposure. But, as impressive as this new toolbox is, it's not the true GIS revolution; the real impact concerns the way we do business. GIS provides the means for the U.S. Forest Service's New Perspective on Forestry Initiative—a capability for consensus building and conflict resolution.

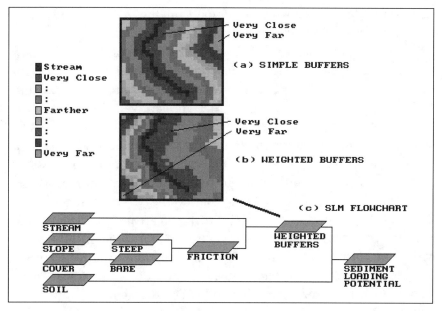

Fig. 14.1. Effective sediment loading map.

All GIS models, whether simple or complex, have three characteristics. They are flexible, succinct, and dynamic. Once a model has been developed, it encourages the addition of new considerations and parameter weights. It almost shouts, "You want to try it a different way?" When was the last time your draftsperson demonstrated such flexibility?

Yet computers, by nature, are stupid. Though they can do a lot of things, they don't know what to do. Each step has to be made perfectly clear. This can be frustrating, but it's valuable in the long run. The GIS model becomes a clear statement of the analysis procedure. It succinctly summarizes the voluminous appendices of most reports. Note the flowchart at the bottom of figure 14.1; it shows that the effective distance from streams is a function of the intervening slope and cover conditions. The flowchart takes sediment loading a step further by considering the following soil conditions: $SLP = fn$ (slope, cover, soils). In other words, you are likely to disturb the dirt if you run your skidder on unstable soils. If this disturbance is effectively close to the stream, you've got a problem. If it's effectively far away (even though it may be "ruler" close), the SLP is low. This is common sense, expressed in five simple sentences to the computer. The model encapsulates and demonstrates your rational thinking. Now the judge and anglers can see what you're talking about.

Finally, GIS models are dynamic. They allow you to try different scenarios, different perspectives. Suppose your model for harvest planning considers visual exposure as well as slope and proximity to roads. The best parcels to log are those that are gently sloped, with good access and minimal visual exposure. What if it is suggested that visual exposure is 10 times more important than the engineering considerations? What parcels, if any, are no longer appropriate for harvesting? Different perspectives might violently disagree in philosophical space. Every square foot may be contested. But do they disagree in geographic space? Where do they disagree? Which parcels are involved? Spatial answers, not ideological statements, are needed. You run the model with your perspective, and I will run it with mine. Subtract the two "solution" maps, and we will see where and how we disagree. That's information for conflict resolution. That's dialogue for consensus building. That's the GIS revolution.

So what's new in natural resources? The seemingly unnatural technology of GIS is what's new. To old foresters, the technology at first appears to cramp their management styles. Like Audrey II in *Little Shop of Horrors*, GIS just keeps growing as it shouts, "Feed me, feed me, *feed me!*" Digitize, digitize, *digitize!* But it's not just a new layer of record keeping. It's more than mapping and spatial data management. It's a toolbox that allows you to plan more realistically. In fact, it's a whole new perspective on the resource decision-making environment. I bet you can learn a new trick or two.

Frankly, My Dear, I Don't Give a Damn

(GIS Facilitates Dialogue and Participatory Involvement)

❖   ❖   ❖

Like Rhett said to Scarlett in *Gone with the Wind*, you may not give a damn. All this new technology is more trouble than it is worth—just like Scarlett. But there is still that spark that attracts you. Maybe, just maybe, there is something to all this GIS infatuation. You have heard bits and pieces of its proclaimed capabilities, how it will change the way you do things. But it certainly is a strange beast, not unlike a platypus. There are bits and pieces of disciplines that are familiar, like computers and drafting and statistics; but when assembled in its strange way, it is hard to understand how it keeps afloat. That's GIS, a strange but compelling beast.

As a technology, I can take it or leave it. GIS is interesting, but it often seems academic or even esoteric. It's just technology for technology's sake, isn't it? Does it really improve decision making or just the vendor's bottom line? A real question. A real concern for "real" folks.

Consider our current policy formation and decision-making environments. Professor Evan Vlachos, a natural resources futurist at Colorado State University, identifies three inputs to this process: the scientific, the social, and the legal components of society. Loosely paraphrased, he says that a delicate balance of all three is necessary to reach an effective decision. For example, assume the populace of a developing country is suffering from a dietary deficiency in animal protein. Also, assume the country has vast areas of natural grasslands. Common sense (and voluminous tables of research from the beef council) point to the easy technical solution of cattle production.

That's it, an obvious, technically supportable solution. But wait a moment. There are indigenous cultural and religious taboos against killing cows. And the legal concept of private property is nonexistent. Nor is there a precedent for land ownership and fencing. In short, the plan receives high technical marks and appears acceptable from that single perspective. But it fails the social and legal tests. A decision bust.

In a less obvious fashion, the necessary balance among the three decision sectors is what upsets our current decision making. A technical solution often meets social acceptance in a loggerhead. The result is an injunction, and the legal sector is called to make the management decision. Black robes and a litany of conflicting testimony replace effective science and social inputs.

Facing the real prospect of litigation, we resort to a state of "analysis paralysis." We constantly search for one more decimal point of accuracy to the technically perfect solution. "If we could just get more data, the answer would be obvious," we say. Don't manage the woods, inventory them. It's like placing a sign at park headquarters, "closed for inventory."

But maybe your data and analysis are good enough—not perfect but good enough—to make a decision. The problem could be the decision environment. What is lacking is a capability to clearly communicate rational thinking and different perspectives. Figure 15.1 summarizes these thoughts. An inability to communicate causes the social and scientific inputs to turn toward the legal sector to solve their perceived differences—litigation. If I am not part of the analysis process, chances are I don't fully understand or trust it. Another hundred pages of appendices or even more colorful maps won't help. See you in court.

However, effective communication causes these inputs to turn away from the legal sector in search of an acceptable decision. The right side of the figure addresses the working environment. Most geographic-related decisions have a relatively broad range of technically feasible options. As the pyramid depicts, there is a smaller set of options that are economically viable. We have developed elaborate procedures for assessing management options that are both feasible and viable—the technical solution.

Yet, in reality, there is an even smaller set of options that are socially acceptable. It is this final sieve of management alternatives that often confounds our decision making. It uses elusive measures such as human values, attitudes, beliefs, judgment, trust, and understanding, but not the usual quantitative stuff used in computer algorithms and decision-making models. So what does all this have to do

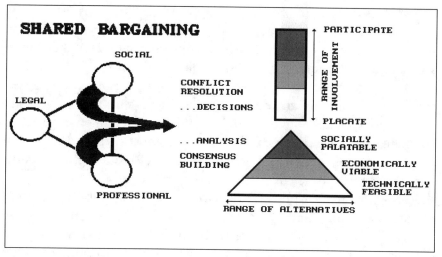

Fig. 15.1. Considerations in effective decision making.

with GIS? In a GIS, maps are numbers, aren't they? GIS is a technology, isn't it? Well, yes and no.

The step from feasible to acceptable options is not so much science and economics as it is communication. And effective communication implies involvement. The range of involvement (upper-right portion of fig. 15.1) extends from placation to actual participation. Public hearings often are more placation than participation. At an initial hearing, concerned parties are asked for their input. But for practical reasons, they are excluded from the analysis process. At a final hearing, they are expected to approve the results of analysis they do not understand. "Trust us," they're told. "Everything is there. Just choose alternative A or B."

Participatory decision making has two main thrusts: consensus building and conflict resolution. Consensus building involves technologically driven communication and occurs during the alternative formation phase of the decision-making process. It involves the GIS specialist's translation of the various considerations raised by a decision team into a spatial model. Once completed, the model is run under a wide variety of conditions and the differences in outcome are noted. From this perspective, any single map solution isn't important. How maps change as different scenarios are tried becomes information with which to make a decision. "What if avoidance of visual exposure is more important than avoidance of steep slopes in siting a new road?

Where does the proposed route change, if at all?" In nearly 20 years of GIS consulting, I have consistently found that seemingly divergent philosophical views most often result in only slightly different map views. This realization often leads to group consensus.

If an agreement can't be negotiated, conflict resolution is necessary. This socially driven communication occurs during the decision-formulation phase. It involves the creation of a "conflicts map" that compares the outcomes from two or more competing uses. Each management parcel (vector polygon or raster cell) is assigned a code describing the conflict over the location. "This location is ideal for preservation, recreation, and development. What should we do?" Make a decision, that's what you need to do. This process often involves human rationalizing or trade-offs. "Look here, I will let you develop this parcel if you agree to assign that one to preservation." The dialogue is far from a mathematical optimization, but often closer to an effective decision. It uses the GIS to focus discussion away from broad ideological positions to the specific project area and its unique distribution of possible uses.

GIS is a mapping tool, a spatial database management technology and an analytic revolution. But what may be its most important attribute is its ability to communicate. It isn't just a better technical answer; it is a more comprehensible one. It fosters discussion that often leads to understanding and, ultimately, to effective decisions. Think of it. All Rhett and Scarlett needed was a little more constructive discussion, and they could have lived happily ever after in a small duplex outside Atlanta.

(Note: See topic 10 for a detailed discussion of using GIS in consensus building and conflict resolution.)

Recommended Reading

Books

Antenucci, J.C. "Planning Applications Demand GIS Functionality." *1993 International GIS Sourcebook*, 204-5. Fort Collins, CO: GIS World, 1992.

Aronoff, S. "What Is a Geographic Information System?" Chapt. 2 in *Geographic Information Systems: A Management Perspective*. Ottawa, Canada: WDL Publications, 1989.

Berry, J. K. "Fundamental Operations in Computer-Assisted Map Analysis." In *Fundamentals of Geographic Information Systems: A Compendium*, ed. W. Ripple, 81-98. Bethesda, MD: American Society of Photogrammetry and Remote Sensing, 1989.

Carter, J.R "On Defining the Geographic Information Systems." In *Fundamentals of Geographic Information Systems: A Compendium*, ed. W. Ripple, 3-7. Bethesda, MD: American Society of Photogrammetry and Remote Sensing, 1989.

Coppock, J., and D. Rhind. "The History of GIS." In *Geographical Information Systems: Principles and Applications*, ed. D.J. Maguire, M.F. Goodchild, and D.W. Rhind, Vol. 1, 45-54. Essex, UK: Longman, 1991.

Dobson, J.E. "Exploring Geographical Analysis." *1990-91 International GIS Sourcebook*, 335-37. Fort Collins, CO: GIS World, 1991.

———. "GIS Applications and Functional Requirements." *1993 International GIS Sourcebook*, 203. Fort Collins, CO: GIS World, 1992.

Goodchild, M.F. "A Taxonomy of GIS Systems." *1990-91 International GIS Sourcebook*, 380-83. Fort Collins, CO: GIS World, 1991.

———. "Toward Enumeration and Classification of GIS Functions." *The GIS Sourcebook*, 22-26. Fort Collins, CO: GIS World, 1989.

Maguire, D.J. "An Overview and Definition of GIS." In *Geographical Information Systems: Principles and Applications*, ed. D.J. Maguire, M.F. Goodchild, and D.W. Rhind, Vol. 1, 9-20. Essex, UK: Longman, 1991.

Parker, H.D. "GIS Concepts." *The GIS Sourcebook*, 1-9. Fort Collins, CO: GIS World, 1989.

Robinove, C.J. "Principles of Logic and the Use of Digital Geographic Information Systems." In *Fundamentals of Geographic Information Systems: A Compendium*, ed. W. Ripple, 61-79. Bethesda, MD: American Society of Photogrammetry and Remote Sensing, 1989.

Smith, T., et al. "Requirements and Principles for the Implementation and Construction of Large-Scale Geographic Information Systems." In *Fundamentals of Geographic Information Systems: A Compendium*, ed. W. Ripple, 19-37. Bethesda, MD: American Society of Photogrammetry and Remote Sensing, 1989.

Star, J., and J. Estes. "Introduction," "Background and History," and "The Essential Elements of a GIS: An Overview." Chapts. 1, 2, and 3 in *Geographic Information Systems: An Introduction*, Englewood Cliffs, NJ: Prentice Hall, 1990.

Tomlin, C.D. "Data Processing" and "Data-Processing Control." Chapts. 2 and 3 in *Geographic Information Systems and Cartographic Modeling*. Englewood Cliffs, NJ: Prentice Hall, 1990.

Unwin, D. *Introductory Spatial Analysis*. London: Methuen, 1981.

Journal Articles

Cowen, D.J. "GIS Versus CAD Versus DBMS: What Are the Differences?" *Photogrammetric Engineering and Remote Sensing* 54: 1551 (1988).

Goodchild, M.F. "A Spatial Analytical Perspective on Geographical Information Systems." *International Journal of Geographical Information Systems* 1(4): 327-34 (1987).

Lam, N.S. "Spatial Interpolation Methods: A Review." *American Cartographer* 10: 129-49 (1983).

Piwowar, J., and E. LeDrew. "Integrating Spatial Data: A User's Perspective." *Photogrammetric Engineering and Remote Sensing* 56(11): 1497-1502 (1990).

Wikle, T.A. "Computers, Maps, and Geographic Information Systems." *National Forum* (Summer 1991): 41-43.

TOPIC 5

ASSESSING VARIABILITY, SHAPE, AND PATTERN OF MAP FEATURES

Make strategies fit the situation; don't try to fit the situation to the strategy.

—*General George C. Patton*

The shape and pattern of landscape features are readily apparent to the eye but historically difficult to quantify. This section describes several indices used in the characterization of the configuration and arrangement of features. Discussion of the assumptions, approaches, and algorithms for each technique is made.

16 | Need for Right Questions Takes You Beyond Mapping

(Indices of Map Variability: Neighborhood Complexity and Comparison)

❖   ❖   ❖

[W]here up so floating, many bells down....
—T.S. Eliot

Is some of this beyond mapping discussion a bit dense? Like a T.S. Eliot poem it is full of significance (?) but somewhat confusing for the uninitiated. I am sure many of you have been left musing, "So what? This GIS processing just sounds like a bunch of gibberish to me." You're right. You are a decision maker, not a technician. The specifics of processing are not beyond you and your familiar map, but such details are best left to the technologists. Or are they?

This concern is the focus of topic 4 where GIS is established as, above all else, a communication device facilitating the discussion and evaluation of different perspectives of our actions on the landscape. The most difficult part of GIS is not digitizing, creating databases, or even communicating with the blasted system. Those are technical considerations with technical solutions outlined in the manual. The most difficult part of GIS is asking the right questions. Those questions involve conceptual considerations requiring you to think spatially. That's why you, the GIS user, need to go beyond mapping so that you can formulate your complex questions about geographic space in a manner the technology can use. GIS can do a lot of things, but it doesn't know what to do without your help. A prerequisite to this partnership is your responsibility to develop an understanding of what GIS can't do.

With this flourish in mind, let's complete our techy discussion of neighborhood operators (started in topic 3). Recall that these techniques involve summarizing the information found in the general vicinity of each map location. These summaries can characterize the surface configuration (e.g., slope and aspect) or generate a statistic (e.g., total and average values). The neighborhood definition, or roving window, can have a simple geometric shape (e.g., all locations within a quarter of a mile) or a complex shape (all locations within a 10-minute drive). Window shape and summary technique define the wealth of neighborhood operators, from simple statistics to spatial derivative and interpolation. So much for review, on to new stuff.

An interesting group of these operators are referred to as *filters*. Most are simple binary or weighted windows as discussed previously. But one has captivated my imagination since Dennis Murphy of the EROS Data Center introduced me to it in the late 1970s. He identified a technique for estimating neighborhood variability of nominal-scale data using a binary comparison matrix (BCM). That's a mouthful of nomenclature, but it's a fairly simple and extremely useful concept. As we become more aware, variability within a landscape plays a significant role in how we (and our other biotic friends) perceive an area. But how can we assess such an illusive concept in decision terms?

Neighborhood Variability

Neighborhood variability can be described two ways, the complexity of an entire neighborhood and the comparison of the conditions within the neighborhood. These concepts can be outlined as follows:

- Complexity (Entire neighborhood)

  –Diversity (Number of different classes)

  –Interspersion (Frequency of class occurrence)

  –Juxtaposition (Spatial arrangement of classes)

- Comparison (Individual vs. neighbors)

  –Proportion (Number of neighbors having the same class as the window center)

  –Deviation (Difference between the window center and the average of its neighbors)

Consider the 3-x-3 window in figure 16.1. Assume M is one class of vegetation (or soil or land use) and F is another. The simplest summary of neighborhood variability is to say there are two classes. If the window had only one class, you would say there is no variability. If it had nine classes, you would say there is a lot of variability. The number of different classes is called *diversity*, the broadest measure of neighborhood variability. If there was only one cell of M and eight of F, you would probably say, "Sure the diversity is still two, but there is less variability than the three of M versus six of F condition in our example." The measure of the frequency of occurrence of each class, termed *interspersion*, is a refinement of the simple diversity count. But doesn't the positioning of the different classes contribute to window variability? It sure does. If the three M's were spread out like a checkerboard, you would probably say there was more variability. The relative positioning of the classes is termed juxtapositioning.

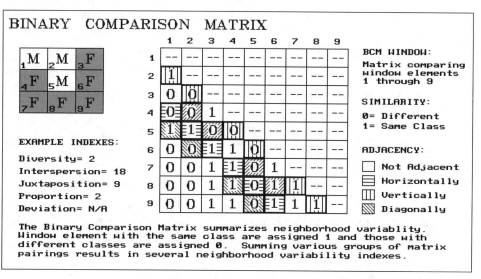

Fig. 16.1. Binary comparison matrix summary of neighborhood variability.

We're not done yet. Neighborhood variability has another dimension. The measures of diversity, interspersion, and juxtapositioning summarize an entire neighborhood's complexity. Another way to view variability is to compare one neighborhood element to its surrounding elements. These measures focus on how different a specific cell is to its surroundings (often termed *anomaly detection*). For our example, we could calculate the number of neighbors with the same classification as the center element. This technique, termed *proportion*, is appropriate for nominal, discontinuous mapped data such as vegetation type maps. For gradient data, as with elevation, *deviation* can be computed by subtracting the average of the neighbors from the center element. The greater the difference, the more unusual the center. The sign of the difference tells you the nature of the anomaly—unusually bigger (+) or smaller (-).

Quantifying Variability

That's a lot of detail. And, like T.S.'s poems, it may seem like a lot of gibberish. You just look at landscape and intuitively sense the degree of variability. Yep, you're smart, but the computer is dumb. It has to quantify the concept of variability. So how does it do it? The binary comparison matrix (BCM), of course. First, *binary* means we only work with zeros and 1s. *Comparison* says we compare each element in the window with every other element. If they are the same, assign a 1. If different, assign a zero. Matrix tells us how the data will be organized.

Now let's put it all together. In figure 16.1, the window elements are numbered from one through nine. Is the class for element one the same as for element two? Yes, so assign a 1 at the top of column one in the table. How about elements one and three? Nope, so assign a zero in the second position of column one. How about one and four? Nope, then assign another zero and so forth, until all columns in the matrix contain a zero or a one. But you are already bored. That's the beauty of the computer. It enjoys completing the table. And yet another table for the next position as the window moves to the right, and the next, and the next, until it has done it thousands of times, roving the window throughout the map.

So why put your silicon subordinate through all this work? Surely its electronics get enough exercise just reading your electronic mail. The work is worth it because the BCM contains the necessary data to quantify variability. It's how your computer sees landscape variability from its digital world. As the computer compares the window elements, it keeps track of the number of different classes it encounters— diversity equals 2. Within the table are 36 possible comparisons. In our example, we find that 18 of these are similar by summing the entire matrix—interspersion equals 18. The relative positioning of classes in the window can be summarized in several ways. Orthogonal adjacency (horizontal and vertical) frequently is used and is computed by summing the highlighted numbers in the table—juxtaposition equals 9. Diagonally adjacent and nonadjacent variability indices sum different sets of window elements. Comparison of the center to its neighbors computes the sum for all pairs involving element five—proportion equals 2.

The techy reader is, by now, bursting with ideas of other ways to summarize the table. The rest of you are back to asking, "So what? Why should I care?" You can ignore the mechanics of the computation and still be a good decision maker. But can you ignore the indices? Sure, if you are willing to visit every hectare of your management area or visually assess every square millimeter of your map. And convince me, your clients, and the judge of your exceptional mental capacity for detail. Or you could learn, on your terms, to interpret the computer's packaging of variability. Does the spotted owl prefer higher or lower juxtapositioning values? What about the pine martin? Or Dan Martin, my neighbor? Extracting meaning from T.S. Eliot is a lot of work. The same goes for unfamiliar analytical capabilities such as the BCM. It's not beyond you. You just need a good reason to take the plunge.

You Can't See the Forest for the Trees

*(Indices of Feature Shape: Boundary
Configuration and Spatial Integrity)*

❖   ❖   ❖

On the other hand, you can't see the trees
for the forest.

—Proverb

The previous article described how the computer sees landscape variability by computing indices of neighborhood "complexity and comparison." This may have incited your spirited reaction, "That's interesting. But, so what. I can see the variability of landscapes at a glance." That's the point. You see it as an image, the computer must calculate it from mapped data. You and your sickly, gray-toned companion live in different worlds—inked lines, colors, and planimeters for you; numbers, algorithms, and "mapematics" for your computer. Can such a marriage last? It's like a hippo and hummingbird romance—bound to go flat.

In the image world of your map, your eye jumps around at what IBM futurist Walter Doherty calls "human viewing speed," or the very fast random access of information. The computer, however, is more methodical. It plods through thousands of serial summaries developed by focusing on each piece of the landscape puzzle. In short, you see the forest; it sees the trees. You couldn't be farther apart. Right?

No, it's just the opposite. The match couldn't be better. Both the strategic and the tactical perspectives are needed for complete understanding of maps. Our cognitive analyses have been fine-tuned through years of experience, but they are hard to summarize and fold into on-the-ground decisions. In the past, our numerical analyses have been as overly simplified as they have been tedious. There's just too much information for human serial processing at the "trees" level of detail. That's where the computer's indices of spatial patterns come in. They provide an entirely new view of your landscape, one that requires understanding and interpretation before it can be used effectively for decision making.

In addition to landscape variability discussed in article 16, the size and shape of individual features affect your impression of spatial patterns. For example, suppose you are a wildlife manager assessing ruffed grouse habitat and population dynamics. The total acreage of suitable habitat is the major determinant of population size. That's a task for the "electronic planimeter" of the GIS toolbox—cell counts in raster systems and table summaries in most vector systems. But is that enough? Likely not, if you want fat and happy birds.

The shape of each habitat unit plays a part. Within a broad context, shape involves two characteristics, boundary configuration and spatial integrity. Consider the top portion of figure 17.1. Both habitat units are 30 acres. Therefore, they should support the same grouse grouping. Right? But research has shown that the birds prefer lots of forest/opening edges. That's the case on the right; it's boring and regular on the left. But what happens if your map has hundreds of individual parcels? Your mind is quickly lost in the "tree" level detail of the "forest."

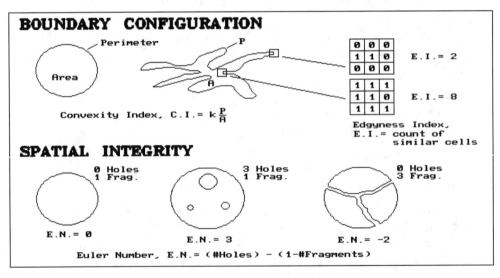

Fig. 17.1. Characterizing boundary configuration and spatial integrity.

Boundary Configuration

That's where the computer comes in. The boundary configuration, or "outward contour," of each feature is calculated as a ratio of the perimeter to the area. In planimetric space, the circle has the least amount of perimeter per unit area. Any other shape has more perimeter and, as a result, a different "convexity index." In the few GISs having this capability, the index uses a "fudge" factor (k) to produce a range of values from 1 to 99. A theoretical zero indicates an infinitely large perimeter around an infinitesimally small area. At the other end, an index of 100 is interpreted as being 100 percent similar to a perfect circle. Values in between define a continuum of boundary regularity. As a GIS user, your challenge is to translate this index into decision terms. "Oh, so the ruffed grouse likes it rough. Then the parcels with convexity indices less than 50 are particularly good, provided they are more than 10 acres, of course." Now you're beyond mapping and actually GISing.

But what about the character of the edge as we move along the boundary of habitat parcels? Are some places better than others? Try an "edginess" count. It's similar to the binary comparison matrix (BCM) discussed in the previous article. A 3-x-3 analysis window is moved about the edge of a map feature. A one is assigned to cells with the same classification as the edge cell; a zero to those that are different. Two extreme results are shown in figure 17.1. A two count indicates an edge location that's really hanging out there. An eight count is an edge, but it is barely exposed to the outside. Which condition does the grouse prefer? Or an elk? Or the members of the Elks Lodge, for that matter? Maybe the factors of your decision making don't care. At least it's comforting to know that such edginess can be quantified in a way the computer can "see" it and spatial modelers can use it.

Spatial Integrity

That brings us to our final consideration—spatial integrity. It involves counting "holes" and "fragments" associated with map features. If a parcel is just one glob, without holes in it, it is said to be intact, or spatially balanced. If holes begin to violate its interior, or it is broken into pieces, the parcel's character changes. Your eye easily assesses that. It is thought that the spotted owl's eye easily assesses that with the bird preferring large, uninterrupted old-growth forest canopies. But how about the computer's eye?

In its digital way, the computer counts the number of holes and fragments for the map features you specify. In a raster system, the algorithms performing the task are fairly involved. In a vector system, the topological structure of the data plays a big part in the processing. That's the programmer's concern. Our concern is understanding what it all means and how we might use it.

The simple counts of holes and fragments are useful data. But these data taken alone can be as misleading as total acreage calculations. Their interplay provides additional information, summarized by the Euler number depicted in the figure. This index tracks the balance between the two elements of spatial integrity by computing their difference. If E.N.=0, the feature is balanced. As you poke more holes in a feature, the index becomes positively unbalanced (large positive values). If you break it into a bunch of pieces, its index becomes negatively unbalanced (large negative values). If you poke it with the same number of holes as you break it into pieces, a feature becomes spatially balanced.

"What? That's gibberish." No, it's actually good information. It can answer such enduring questions as "Does a zebra have white stripes on a black backgound or black stripes on a white background?" Or, "Is an area best characterized as urban pockets surrounded by a natural landscape, or as natural areas surrounded by urban sprawl?" Or, "As we continue clear-cutting the forest, when do we change the fabric of the landscape from a forest with clear-cut patches to islands of trees within a clear-cut backdrop?" It's more than simple area calculations of the GIS.

Shape analysis is more than a simple impression you get as you look at a map. It's more than tabular descriptions in a map's legend. It's both the "forest" and the "trees," an informational interplay between your reasoning and the computer's calculations.

Discovering Feature Patterns

(Assessing Landscape Pattern: Spacing and Contiguity)

❖   ❖   ❖

Everything has its place; everything in its place.
—Granny

Granny was as insightful as she was practical. Her prodding to get socks picked up and in the drawer is actually a lesson in basic ecology. The results of dynamic interactions within a complex web of physical and biological factors puts "everything in its place." The obvious outcome of this process is the unique arrangement of land-cover features that seem to be tossed across a landscape. Such a seemingly disorganized arrangement is nurtured by Mother Nature. It's good she never met Granny.

Articles 16 and 17 deal with quantifying spatial arrangements into landscape variability and individual feature shape. Another characteristic your eye senses as you view a landscape is the pattern formed by the collection of individual features. People often use terms such as *dispersed* or *diffused* and *bunched* or *clumped* to describe patterns formed on the landscape. However, these terms are useless to our "senseless" computer. It doesn't see the landscape as an image nor has it had the years of practical experience required for such judgment. Terms describing patterns are visceral. You just know these things. Stupid computer; it hasn't a clue. Or has it?

As established in articles 16 and 17, the computer "sees" the landscape in an entirely different way—digitally. Its view isn't a continuum of colors and shadings that form features but an overwhelming pile of numbers. The real difference is that you use "experience" and it uses "computation" to sort through the spatial information.

So how does a computer analyze a pattern formed by the collection of map features? The computer's view of landscape patterns must be some sort of mathematical summary of numbers. Over the years, a wealth of indices have been suggested. Most of the measures can be divided into two broad approaches: those summarizing individual feature characteristics and those summarizing spacing among features.

Feature Characteristics

Feature characteristics such as abundance, size, and shape can be summarized for an entire landscape. These landscape statistics provide a glimpse of overall feature patterns. Imagine a large, forested area pocketed with clearcut patches. A simple count of the number of clearcuts gives you a "first cut" measure of forest fragmentation. An area with hundreds of cuts is likely to be more fragmented than an equal-sized area with only a few. But it also depends on the size of each cut and, as discussed in article 17, the shape of each cut.

Putting size and shape together over an entire area is the basis of fractal geometry. In mathematical terms, the fractal dimension, D, is used to quantify the complexity of the shape of features using a perimeter-area relation, specifically, $P \approx A^{**}(D/2)$, where P is the patch perimeter and A is the patch area. The fractal dimension for an entire area is estimated by regressing the logarithm of a patch area on its corresponding log-transformed perimeter. Whew! Imposing mathematical mechanics, but a fairly simple concept; more edge for a given area of patches means things are more complex. To the user, it is sufficient to know that the fractal dimension is simply a useful index. As it gets larger, it indicates an increasing departure from Euclidean geometry. Or, in more human terms, a large index indicates a more fragmented forest and, quite possibly, more irritable beasts and birds.

Feature Spacing

Feature spacing addresses another landscape pattern aspect. With a ruler, you can measure the distance from the center of each clearcut patch to the center of its nearest neighboring patch. The average of all the nearest-neighbor distances characterizes feature spacing for an entire landscape. This is theoretically simple, but too tedious to implement and too generalized to be useful. It works great on scattering marbles. But as patch size and density increase and shapes become more irregular, this measure of feature spacing becomes ineffective. The merging of both area-perimeter characterization and nearest-neighbor spacing into an index provides much better estimates.

For example, dispersion, a frequently used measure developed in the 1950s, uses the following equation: $R = 2((p**.5)*r)$, where R is dispersion, r is the average nearest neighbor distance and p is the average patch density (computed as the number of patches per unit area). When R equals 1, a completely random patch requirement is indicated. A dispersion value less than 1 indicates a more regular dispersed pattern.

All the equations, however, are based in scaler mathematics and simply use GIS to calculate equation parameters. This is not a step beyond mapping but an automation of current practice. Consider the right-hand side of figure 18.1 for a couple of new approaches. The center two plots depict two radically different patterns of "globs," a systematic arrangement (pattern A) on the top and an aggregated one on the bottom (pattern B).

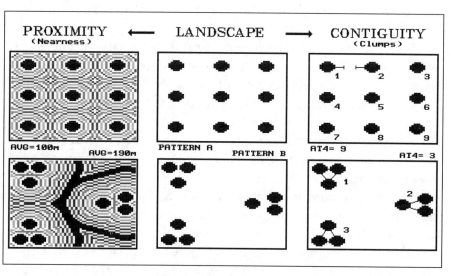

Fig. 18.1. Characterizing map feature spacing and pattern.

The proximity measure on the left side forms a continuous surface of buffers around each glob. The result is a proximity surface indicating the distance from each map location to its nearest glob. For the systematic pattern A, the average proximity is only 100 meters (five "steps" of 20 meters each), with a maximum distance of 220 meters and a standard deviation of ±40 meters. The aggregated pattern B has a much larger average of 190 meters, with a maximum distance of 500 meters and a larger standard deviation of ±120 meters. Where the broad band starts, it's more than 300 meters to the nearest glob, much more than the farthest distance in the systematic pattern. Your eye senses this void; the computer recognizes it as having large proximity values.

The contiguity measure on the right side of the figure takes a different perspective. It looks at how the globs are grouped. It asks the question, "If each glob is allowed to reach out a bit, which ones are so close they will effectively touch?" If the "reach at" factor is only one "step" of 20 meters, none of the nine individual clumps will be grouped in either pattern A or B. However, if the factor is 2, some grouping occurs in pattern B, and the total number of extended clumps are reduced to six. As shown in figure 18.1, an "at" factor of 4 results in just three extended clumps for the aggregated pattern. The systematic pattern is still left with the original nine. Your eye senses the nearness of globs; the computer recognizes this same thing as effective clump numbers.

See? Both you and your computer can see the pattern differences. But the computer sees it in a quantitative fashion with a lot more detail in its summaries. Instead of just a simple average proximity between globs, it generates a distribution of feature spacing—100 percent of pattern A is 11 steps or less away; only 67 percent of pattern B is 11 steps or less away. The computer can describe the distribution of feature spacing as either a cumulative frequency table or a map image —either quantitatively or geographically.

But there is more. Remember articles 4, 5, and 6 that describe *effective distance*? Not all things align themselves in straight lines "as the crow flies." Suppose some patches are separated by streams your beast of interest can't cross or areas with high human activity that they could cross, but prefer not to cross unless they have to. Now, what is the real feature spacing? You don't have a clue. But the proximity and contiguity distributions will tell you what it is really like to move among the features. Without the computer, you must assume your animal moves in the straight line of a ruler, and the real-world complexity of landscape patterns can be reduced to a single value. These are bold assumptions that ask little of GIS. To go beyond mapping, GIS asks a great deal of you—to rethink your assumptions and methodology in light of its new tools.

Recommended Reading

Books

Abler, R., J. Adams, and P. Gould. *Spatial Organization: The Geographer's View of the World*. Englewood Cliffs, NJ: Prentice Hall, 1971.

Buttenfield, B., and W. Mackaness. "Visualization." In *Geographical Information Systems: Principles and Applications*, ed. D.J. Maguire, M.F. Goodchild, and D.W. Rhind, Vol. 1, 427-43. Essex, UK: Longman, 1991.

Mandelbrot, B.B. *The Fractal Geometry of Nature*. San Francisco, CA: Freeman Press, 1982.

Munkres, J.R. *Topology: A First Course*. Englewood Cliffs, NJ: Prentice Hall, 1975.

Tomlin, C.D. "Data." Chapt. 1 in *Geographic Information Systems and Cartographic Modeling*. Englewood Cliffs, NJ: Prentice Hall, 1990.

Turner, M.G., ed. *Landscape Heterogeneity and Disturbance*. New York: Springer-Verlag, 1987.

Journal Articles

Baker, W.L. "A Review of Models of Landscape Change." *Landscape Ecology* 2: 111-33 (1989).

Forman, R., and M. Gordon. "Patches and Structural Components for Landscape Ecology." *BioScience* 31: 733-40 (1981).

LaGro, J.A. "Assessing Patch Shape in Landscape Mosaics." *Photogrammetric Engineering and Remote Sensing* 57: 285-93 (1991).

Milne, B.T. "Measuring the Fractal Geometry of Landscapes." *Applied Mathematics and Computation* 27: 67-79 (1988).

O'Niell et al. "Indices of Landscape Pattern." *Landscape Ecology* 1(3): 153-62 (1988).

Ripple et al. "Measuring Forest Landscape Patterns in the Cascade Range of Oregon." *Biological Conservation* 57: 73-88 (1991).

Turner, M.G. "Landscape Ecology: The Effect of Pattern on Process." *Annual Review of Ecology and Systematics* 20: 171-97 (1991).

TOPIC 6

OVERLAYING MAPS AND CHARACTERIZING ERROR PROPAGATION

Perfect is the enemy of good.

—Anonymous

Overlaying maps is at the heart of most GIS applications. However, the propagation of errors needs to be characterized and included with the overlay results. This section describes approaches used in establishing map uncertainty and assessing error propagation.

GIS Facilitates Error Assessment

(Creating a Shadow Map of Uncertainty: Probability Surface)

❖   ❖   ❖

Q: Cross an elephant with a rhinoceros,
       and what do you get?

A: El-if-I-know!

Overlay a soils map with a forest-cover map, and what do you get? El-if-I-know! It's supposed to be a map that indicates the soil/forest conditions throughout a project area. But are you sure that's what it is? Where are you sure the reported coincidence is right? Where are you less sure about the results? Stated in the traditional opaque academic way, "What is the spatial distribution of probable error associated with your GIS modeling product?"

It's not enough simply to jam a few maps together and presume the results are inviolately accurate. You have surely heard of the old adage, "Garbage in, garbage out." But even if you have good data as input, the "throughput" can garble your output. Whoa! That's nonsense. As long as one purchases a state-of-the-art system and is careful in constructing the database, everything will be OK. Right? Well, maybe there are a couple of other things to consider, like map uncertainty and error assessment.

Potential Error Sources

There are two broad types of GIS errors: those present in the encoded base maps and those that arise during analysis. The source documents you encode may be inherently wrong; or your encoding process may introduce error. The result in either case is garbage poised to be converted into more garbage—at megahertz speed and in vibrant colors. But things aren't that simple, just good or bad data, right or wrong. The ability of the interpretation process to characterize a map feature comes into play.

Things always seem a bit more complicated in a GIS. Traditional scaler mathematical models reside in numeric space, not in the seemingly chaotic reality of geographic space. Spatial data have both a "what" (thematic attribute) and a "where" (locational attribute). These are two avenues to error. Consider a class of aspiring photointerpretation students. Some students will outline a tightly clustered stand of trees and mark it as ponderosa pine. Others will extend their boundaries to encompass a few nearby trees scattered in the adjoining meadow and similarly mark the stand as ponderosa pine. The remainder of the class will trace a different set of boundaries and mark their renderings as lodgepole pine. Whose feature definition are you going to treat as gospel in your GIS?

Unless a feature exhibits a sharp edge and is properly surveyed, there is a strong chance the boundary line is misplaced in at least a few locations. Even exhibiting a sharp edge may not be enough. Consider a lake on an aerial photograph. Using your finest tipped pen and drafting skills, you still cannot draw an accurate line. A month later the shoreline might recede 100 yards. Next spring it could be 100 yards in the other direction. So where is the lake? El-if-I-know! But I do know it's somewhere between here (high-water mark) and there (low-water mark). If it's late spring, it's likely to be near here.

A Unique Approach

This puts map error in a new light. Instead of a sharp boundary implying absolute truth, a probability surface can be used. Consider a typical soil map. The polygonal edge implies that soil A stops here, and soil B begins immediately. Like the boundary between you and your neighbor, the transition space is zero and the characteristics of the adjoining features are absolutely known. That's not likely the case for soils; it's more like a probability distribution but expressed in map space.

Imagine a typical soil map with a probability "shadow" map clinging to it—sort of glued to the bottom. Your eye goes to a location on the map and notes the most likely soil from the top map, then peers through to the bottom to see how likely it actually is. For human viewing, you could assign a color to each soil type, just like we do now. Then you could control the "brightness" of the color based on the soil's likelihood—a washed-out pink if you're not sure it's soil A; a deep red if you're certain. That's an interesting map with colors telling you the soil type (just like before), yet the brightness adds information about map uncertainty.

The computer treats map uncertainty similarly; it can store two separate maps (or fields in the attribute table)—one for classification type and another for uncertainty. Or for efficiency's sake, the computer can use a "compound" number with the first two digits containing the classification type and the second two digits its likelihood—sort of a number sandwich that can be peeled apart into the two components identifying the "what" characteristic of maps: What do you think is there, and how sure are you?

An intriguing concept, but is it practical? We have enough trouble preparing traditional maps, let alone a shadow map of probable error. For starters, how do we currently report map error? If map error is reported at all, it's broadly discussed in the map's legend or appended notes. For some maps, errors are field evaluated and an error matrix is reported. This involves locating yourself in a project area, noting what is around you and comparing this with what the map predicts. Do this a few hundred times and you get a good idea of how well the map is performing (e.g., ponderosa pine was correctly classified 80 percent of the time). If you keep track of the errors that occurred, you also know something about which map features are being confused with others (e.g., 10 percent of the time lodgepole pine was incorrectly identified as ponderosa pine).

Good information; it alerts us to the reality of map errors and even describes the confusion among map features. Still, the information offers a spatially aggregated error assessment, not a continuous shadow map of error. However, insight is provided into one of the elements necessary to the development of the shadow map. Recall the example of the students in the photointerpretation course described earlier. There are two ways the students can go wrong: they can imprecisely outline the boundary or they can inaccurately classify the feature. The error matrix summarizes only classification accuracy.

But what about the precision of a boundary's placement? That's a realm that touches on *fuzzy logic*, a new field in mathematics. You might be sure the boundary is around here somewhere, but you may not be sure of its exact placement. If this is the case, you have a transition gradient, not a sharp line. Imagine a digitized line, separating soil A from soil B. Now imagine a series of distance zones about the line.

Right at the boundary line, things are pretty unsure, say 50/50 as to whether it's soil A or B. But as you move away from the line and into the area of soil A, you become more certain it's A. Each of the distance zones can be assigned a slightly higher confidence, say 60 percent for the first zone, 70 percent for the second, etc.

Keep in mind the transition may not be a simple linear one, nor the same for all soils. Also, recall the information in the error matrix likely indicates that we never reach 100 percent certainty. These factors form the ingredients of the shadow map of error. Moreover, they form a challenge to GIS—to conceptualize many map themes in a new, potentially more realistic way. GIS is not just the cartographic translation of existing maps into the computer. The digital map is inherently different from a paper map. The ability to map uncertainty is a big part of this difference. The ability to assess error propagation as we analyze these data is covered in article 20.

20 Analyzing the Nonanalytical

(Error Propagation in GIS Modeling: Joint Probability and Minimum Mapping Resolution)

In the previous article I took the position that most maps contain error, and some maps actually are riddled with it. For most readers, this wasn't so much a revelation as a recognition of reality. Be realistic. A soil or vegetation map is just an estimate of the actual landscape pattern of these features. No one used a transit to survey the boundaries. Heck, we're not even positive the classification is right. Field checking every square inch is out of the question.

Under some conditions, locational and thematic guesses are pretty good; under other conditions, they're pretty bad. In light of this, it is suggested that all maps should contain "truth in labeling," directing a user's attention to both the nature and extent of map uncertainty. You are asked to imagine a typical soil map with a shadow map of probable error glued to its bottom. You could look at the top to see the soil classification, and then peer through to the bottom layer to see how likely the classification is at that location. Error matrices and fuzzy gradients are discussed as procedures for getting a handle on this added information.

Error Propagation

But regardless of how we derive a shadow map of error, how would we use such information? Is it worth all the confusion and effort? In manual map analysis, the interaction of map error is difficult to track and often deemed not worth the hassle. However, digital map analysis is an entirely new ball game. When error propagation comes into play, it takes us beyond mapping and even beyond contemporary GISing. Error propagation extends the concept of cartographic modeling to spatial modeling. Applications can move from planning models to process models. Whoa! This is beginning to sound like an academic beating, a hundred lashes with arcane terminology longer than a bull whip.

Map error propagation identifies a new frontier for GIS technology, error propagation modeling. Suppose you want to overlay vegetation and soil maps to identify areas of Douglas fir and Cohasset soil. But realistically, you're not completely comfortable with either map's accuracy. Even if the vegetation or soil classifications are correct, the boundaries may not be exactly right. The simplest error propagation model is the joint probability model. Let's assume the two soil and vegetation features are barely overlapping, as depicted in the left portion of figure 20.1. That means, if it's only a coin flip that the overlap is Douglas fir (.5), and it's only a coin flip that it's Cohasset soil (.5), then it's a long shot (.5 x .5 = .25) that both are present at that location. In other words, the Douglas fir and Cohasset combination is the best estimate, but don't put too much stock in it. Another map location within the bounds of both features is a lot more certain of the coincidence. Common sense and the joint probability of 1.0 x 1.0 = 1.0 tells us that.

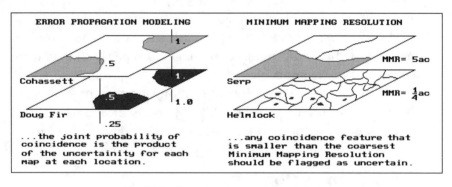

Fig. 20.1. Characterizing map uncertainty.

An even better estimate of propagated error is a *weighted joint probability*. The phrase is a nomenclature mouthful, but it's an easy concept. If you are not sure about a location's classification, but the classification doesn't have much impact on your model, then don't worry about it. However, if your model is sensitive to a map variable and you're not sure of the variable's classification, then don't put much stock in the prediction.

Techy types should immediately recognize that the error propagation weights are the same as the model weights (e.g., the X coefficients in a regression equation). In implementation, you spatially evaluate your equation using the maps as variables. At each map location, a set of values are "pulled" from the stack of maps, the equation is solved, and the result is assigned to that location on the solution map. However, you are not done until you pull error estimates from the set of shadow maps, compute a weighted joint probability, and attach it to the bottom of the solution map. Vendors and users willing, this capacity will be part of the next generation of GIS packages.

More Than Meets the Eye

So where does this leave us? We have, for the most part, translated the basic concepts and procedures of manual map analysis into our GIS packages. Concepts such as *map scale* and *projection* are efficiently accounted for and recorded. Some packages even pop up a warning if you try to overlay two maps with different scales. You can then opt to rescale to a common base before proceeding. But should you? Is a simple geographic scale adjustment sufficient? Is there more to this than meets the eye? The rescaling procedure is mathematically exacting; simply multiply the x and y coordinates by a conversion factor. But the procedure ignores the informational scale implications, namely, thematic error assessment.

A simple classroom exercise illustrates this concern. Students use a jeweler's loupe to measure the line thickness of a stream depicted on both 1:24,000 and 1:100,000 map sheets. The lines are about the same thickness, which implies a stream several feet wide at the larger scale, and one tens of feet wide at the smaller scale. Which measure is correct? Or is flooding implied? Using a photocopier, the students enlarge the 1:100,000 map to match the other one. The two maps are woefully dissimilar. The jigs and jags in one are depicted as a fat, smooth line in the other. Which is correct? Or is stream rechanneling implied?

The discrepancy is the result of mixing scales. The copier adjusted the geographic scale but ignored the informational scale differences. In a GIS, rescaling is done in a blink of the eye, but it should be done with proper reverence. Estimation of induced error should be incorporated in the rescaled product.

Minimum mapping resolution (MMR) is another aspect of informational scale. MMR reports the level of spatial aggregation. All maps have some level of spatial aggregation. And with spatial aggregation comes variation, a real slap in the face to map accuracy. A soil map, for example, is only a true representation at the molecular level. If you aggregate to dust particles, I bet there will be a few stray foreign molecules tossed in. Even a dirt clod map likely will have a few foreign particles tossed in. In a similar light, does one tree constitute a timber stand on a vegetation map? Or does it take two trees? What about a clump of 20 pines with one hemlock in the center? MMR reports the smallest area that can be circled and called one thing.

Further Consideration

So what? Who cares? Consider the right side of figure 20.1. If we overlay the soil and vegetation maps again, and this time identify locations of serpentine soil and hemlocks, an interesting conclusion can be drawn: there are a lot of hemlocks growing in serpentine soils. That's interesting, because foresters tell us that never happens. But there they are.

What is going on? Mixing scales again, that's what. Any interpreter of aerial photography can see individual hemlocks in a sea of deciduous trees in winter imagery. But you don't circle just one, or else you'll be left with just an ink dot. So you circle stands of about a quarter acre, forming small polygons. Soil mapping is often a tougher task. You have to look through the vegetation mask, note subtle topography changes, and mix in a lot of intuition before circling a soil feature. Because serpentine soil features are particularly difficult to detect, a 5-acre polygon is about as small as you can go. However, be sure to place a marginal note in the soil map's legend about frequent alluvial pockets of about a quarter acre in the area.

That's the reality of the serpentine and hemlock coincidence—the trees are growing on alluvial pockets smaller than the MMR of the soil map. But the GIS said it was growing in serpentine soil. That's induced error from mixing informational scales. So what can we do? The simplest approach is to "dissolve" any extremely small polygonal progeny into their surroundings. Another approach "tags" each coincidence feature that is smaller than the coarsest MMR with an error estimate—sort of a warning that you may be wrong.

Whew! Many GISers are content to go with the best guess of map coincidence and have no use for mapping error. A cartographic model that automates manual map analysis procedures is often worth its weight in gold. Yet there is a growing interest in spatial models with all the rights, privileges, and responsibilities of true "map-ematics." A large part of this rigor is the extension of mathematical procedures, such as error assessment, to GIS technology. We are just scratching the surface of this extension. In doing so, we have uncovered a closet of old skeletons defining map content, structure, and use. The digital map and new analytic capabilities are challenging these historical concepts and rapidly redefining the cartographic playing field.

Recommended Reading

Books

Aronoff, S. "Data Quality." Chapt. 6 in *Geographic Information Systems: A Management Perspective*. Ottawa, Canada: WDL Publications, 1989.

Burrough, P.A. "Data Quality, Errors, and Natural Variation." Chapt. 6 in *Principles of Geographical Information Systems for Land Resources Assessment*. Oxford, UK: Oxford University Press, 1987.

Christman, N.R. "The Error Component in Spatial Data." In *Geographical Information Systems: Principles and Applications*, ed. D.J. Maguire, M.F. Goodchild, and D.W. Rhind, Vol. 1, 165-74. Essex, UK: Longman, 1991.

Drummond, J. "A Framework for Handling Error in Geographic Data Manipulation." In *Fundamentals of Geographic Information Systems: A Compendium*, ed. W. Ripple, 109-18. Bethesda, MD: American Society of Photogrammetry and Remote Sensing, 1989.

Fisher, P.F. "Spatial Data Sources and Data Problems." In *Geographical Information Systems: Principles and Applications*, ed. D.J. Maguire, M.F. Goodchild, and D.W. Rhind, Vol. 1, 191-213. Essex, UK: Longman, 1991.

Newcomer, J.A., and J. Szajgin. "Accumulation of Thematic Map Error in Digital Overlay Analysis." In *Fundamentals of Geographic Information Systems: A Compendium*, ed. W. Ripple, 129-33. Bethesda, MD: American Society of Photogrammetry and Remote Sensing, 1989.

Wehde, M. "Grid Cell Size in Relation to Errors in Maps and Inventories Produced by Computerized Map Processing." In *Fundamentals of Geographic Information Systems: A Compendium*, ed. W. Ripple, 119-28. Bethesda, MD: American Society of Photogrammetry and Remote Sensing, 1989.

Journal Articles

Burrough, P.A. "Fuzzy Mathematical Methods for Soil Survey and Land Classification." *Journal of Soil Science* 40: 477-92 (1989).

Chou, Y.H. "Map Resolution and Spatial Autocorrelation." *Geographical Analysis* 23: 228-46 (1991).

Heatwole, C., and T. Zhang. "Representing Uncertainty in Knowledge-Based Systems: Confirmation, Probability and Fuzzy Set Theories." *Transaction of the ASAE* 33: 314-23 (1990).

Story, M., and R. Congalton. "Accuracy Assessment: A User's Perspective." *Photogrammetric Engineering and Remote Sensing* 52(3): 397-99 (1986).

Walsh, S., D. Lightfoot, and D. Butler. "Recognition and Assessment of Error in Geographic Information Systems." *Photogrammetric Engineering and Remote Sensing* 53(10): 1423-30 (1987).

Wiens, J.A. "Spatial Scaling in Ecology." *Functional Ecology* 3: 385-97 (1989).

TOPIC 7

OVERLAYING MAPS AND SUMMARIZING THE RESULTS

GIS is a map that makes a map.

–Anonymous

In GIS overlaying maps goes beyond traditional procedures of "sandwiching" map sheets on a light-table. In this section, procedures for point-by-point, regionwide, and mapwide overlay summaries are described. Numerous applications and the underlying concepts are presented.

21 Characterizing Spatial Coincidence the Computer's Way

(Point-by-Point Map Overlay Techniques)

❖ ❖ ❖

That's the beauty of the pseudosciences. Since they don't depend on empirical verification, anything can be explained.

–Doonesbury

As noted in the previous article, GIS maps are composed of numbers, and a rigorous, quantitative approach to map analysis should be maintained. However, most of our experience with maps is nonquantitative, using map sheets composed of inked lines, shadings, symbology, and zip-a-tone or colored acetate. We rarely think of map uncertainty and error propagation. And we certainly wouldn't think of demanding such capabilities in our GIS software. That is, not yet.

Everybody knows the "bread and butter" of a GIS is its ability to overlay maps. Why, this is one of the first questions we ask a vendor (right after viewing the three-dimensional plot that knocks our socks off). Most often the question and the answer are framed in our common understanding of "light-table gymnastics." We can conceptualize peering through a stack of acetate sheets and interpreting the subtle hues of resulting colors. With a GIS, we're asking the computer to identify a condition from each map layer for every location in a project area. From the computer's perspective, however, this is just one of many ways to characterize the spatial coincidence.

Comparisons and Contrasts

Let's compare how you and your computer identify map coincidence. Your eyes move randomly about the stack of acetate sheets, pausing for a nanosecond at each location and mentally establishing the conditions by interpreting the colors. Your summary might conclude that the northeastern portion of the area is unfavorable as it has "kind of a magenta tone." This color is the result of visually combining steep slopes portrayed as bright red with unstable soils portrayed as bright blue with minimal vegetation portrayed as dark green. If you wanted to express the result in map form, you would tape a clear acetate sheet on top of the stack, delineate globs of color differences and label each parcel with your interpretation. Whew! No wonder you want a GIS.

The GIS accomplishes the task in a similar manner. In a vector system, line segments defining polygon boundaries are tested to determine if they cross. When a line on one map crosses a line on another map, a new combinatorial polygonal is indicated. Trigonometry is used to compute the x,y coordinate of the line intersection. The two line segments are split into four, and values that identify the combined map conditions are assigned. The result of all this crossing and splitting is the set of polygonal progeny you so laboriously delineated by hand.

A raster system has things a bit easier. Because all locations are predefined as a consistent set of cells within a matrix, the computer merely "goes" to a location, retrieves the information stored for each map layer and assigns a value indicating the combined map conditions. The result is a new set of values for the matrix that identifies map coincidence.

The big difference between ocular and computerized approaches to map overlay isn't so much in technique as in data treatment. If you have several maps to overlay, you quickly run out of distinct colors, and the whole stack of maps turns an indistinguishable purplish hue. One remedy is to classify each map layer into two categories, such as suitable and unsuitable. Keep one category as clear acetate (good) and shade the other in light gray (bad). The resulting stack avoids the ambiguities of color combinations and depicts the best areas as lighter tones. However, in making the technique operable, you have severely limited the content of the data—just good and bad.

Computer Mimicry

The computer can mimic this technique by using binary maps. A 0 is assigned to good conditions and a 1 is assigned to bad conditions. The sum of the maps has the same information as the brightness scale you observe—the smaller the value the better. The two basic logical combination forms can be computed. For example,

```
FIND THOSE LOCATIONS THAT HAVE GOOD SLOPES
.AND. GOOD SOILS .AND. GOOD VEGETATIVE COVER.
```

Your eye sees this as the perfectly clear locations. The computer sees this as the numeric pattern 0-0-0.

```
FIND THOSE LOCATIONS THAT HAVE GOOD SLOPES
.OR. GOOD SOILS .OR. GOOD VEGETATIVE COVER.
```

To you this could be any location that is not the darkest shading; to the computer it is any numeric pattern that has at least one 0. But how would you handle,

```
FIND THOSE LOCATIONS THAT HAVE GOOD SLOPES
.OR. GOOD SOILS .AND. GOOD VEGETATIVE COVER?
```

You can't find these sites by simply viewing the stack of maps. You would have to spend a lot of time flipping through the stack. To the computer, this situation is simply the patterns 0-1-0, 1-0-0, and 0-0-0. It's a piece of cake from the digital perspective.

In fact, any combination is easy to identify. Let's expand our informational scale and redefine each map from just good and bad to not suitable (0), poor (1), marginal (2), good (3), and excellent (4). We can ask the computer to

```
INTERSECT SLOPES WITH SOILS WITH COVER
COMPLETELY FOR ALL COMBINATIONS.
```

The result is a map indicating all combinations that actually occur among the three maps. This map is too complex for human viewing enjoyment, but it contains the detailed information basic to many application models. A more direct approach is a geographic search for the best areas invoked by asking the computer to

```
INTERSECT SLOPES WITH SOILS WITH COVER FOR
EXCELLENT AREAS ASSIGNING 1 TO 4 AND 4 AND 4.
```

Any combination not assigned a value drops to zero, leaving a map with ones indicating the excellent areas.

Alternative Approaches

Let's try another way of combining these maps by asking the computer to

```
COMPUTE SLOPES MINIMIZE SOILS MINIMIZE
COVER FOR WEAK LINK.
```

The resulting map values indicate the minimal coincidence rating for each location. Low values indicate areas of concern, and zero indicates areas to dismiss as not suitable from at least one map's information. There are many other computational operations you could invoke such as plus, minus, times, divide, and exponentiate. Just look at the functional keys on your hand calculator. But you may wonder, "Why would someone want to raise one map to the power of another?" Spatial modelers who have gone beyond mapping, that's who.

What would happen if, for each location (be it a polygon or a cell), we computed the sum of the three maps, then divided by the number of maps? This would yield the average rating for each location. Those locations with higher averages are better. Right? You can take this concept a few steps further. First, in a particular application, some maps may be more important than others in determining the best areas. Ask the computer to

AVERAGE SLOPES TIMES 5 WITH SOILS TIMES 3
WITH COVER TIMES 1 FOR WEIGHTED AVERAGE.

The result is a map whose average ratings are more heavily influenced by slope and soil conditions than cover.

To get a handle on the variability of ratings at each location, you can determine the standard deviation, either simple or weighted. Or for even more information, you can determine the coefficient of variation, which is the ratio of the standard deviation to the average expressed as a percent. What will that tell you? It hints at the degree of confidence you should put into the average rating. A high coefficient of variability indicates wildly fluctuating ratings among the maps, and you might want to look at the actual combinations before making a decision.

How about one final consideration? Combine the minimal rating information (WEAKEST LINK) with that of the average rating (WEIGHTED AVERAGE). A prudent decision maker would be interested in those areas with the highest average rating and that score at least 2 (marginal) in any of the map layers. This level of detail should be running through your head while you view a stack of acetate sheets or a simple GIS product depicting map coincidence. Is it? If not, you might consider stepping beyond mapping.

22 Map Overlay Techniques: There's More Than One

(Regionwide Summaries of Map Coincidence)

❖   ❖   ❖

I have the feeling we're not in Kansas
anymore, Toto.
                    —Dorothy, in The Wizard of Oz

In the previous article the discussion of map overlay procedures may have felt like a scene from *The Wizard of Oz*. The simple concept of throwing a couple of maps on a light-table was blown out of proportion into the techy terms of combinatorial, computational, and statistical summaries of map coincidence. An uncomfortable, unfriendly, and unfathomable way of thinking. But that's the reality of GIS—the surrealistic world of "map-ematics."

Now that maps are digital, all GIS processing is the mathematical or statistical summary of map values. What characterized the discussion in article 21 was that the values to be summarized were obtained from a set of spatially registered maps at a particular location; this was termed "point-by-point map overlay." As in the movie *TRON*, imagine you have shrunk small enough to crawl into your computer and find yourself standing atop a stack of maps. You look down and see numbers aligned beneath you. You grab a spear and thrust it straight down into the stack. As you pull it up, the impaled values form a shish kebab of numbers. You run with the kebab to the central processing unit (CPU) and mathematically or statistically summarize the numbers as they are peeled off. Then you run back to the stack, place the summary value where you previously stood, and move over to the next cell in a raster system. Or, if you're using a vector system, you would move over to the next "polygonal progeny."

Exploring the Possibilities

In the previous article, I identified ways to summarize the values. Let's continue with the smorgasbord of possibilities. Consider a *coincidence summary* that identifies the frequency of joint occurrence. If you

```
CROSSTAB FORESTS WITH SOILS,
```

a table results indicating how often each forest type jointly occurs with each soil type. In a vector system, this is the total area of all the polygonal progeny for each of the forest/soil combinations. In a raster system, this is simply a count of all the cell locations for each forest/soil combination.

For example, the first row of table 22.1 shows that forest category 1 (deciduous) contains 303 cells distributed throughout the map. Soils category 1 (lowland) totals 427 cells. The next column shows that the joint condition of deciduous/lowland occurs 299 times for 47.84 percent of the total map area. Contrast this result with the deciduous/upland occurrence in the row below, indicating only four "crosses" for less than 1 percent of the map. The coincidence statistics for the conifer category are more balanced, with 128 cells (20.48 percent) occurring with the lowland soil and 194 cells (31.04 percent) occurring with the upland soil.

Table 22.1. Coincidence table.

Coincidence Table for Map 1 = FORESTS with Map 2 = SOILS							
Map 1 FORESTS	No. of Cells	Map 2 SOILS	No. of Cells	No. of Cross	% of Total	% Map 1 Area	% Map 2 Area
1 Deciduous	303	1 Lowland	427	299	47.84	98.68	70.02
1 Deciduous	303	2 Upland	198	4	0.64	1.32	2.02
2 Conifer	322	1 Lowland	427	128	20.48	39.75	29.98
2 Conifer	322	2 Upland	198	194	31.04	60.25	97.98

These data may cause you to jump to conclusions, but you better consider the last section of the table before you do. This section normalizes the coincidence count to the total number of cells in each category. For example, the 299 deciduous/lowland coincidence accounts for 98.68 percent of all deciduous tree occurrences ((299/303)*100). That's a strong relationship. However, in the lowland soil occurrence, the 299 deciduous/ lowland coincidence count is weaker as it accounts for only 70.02 percent of all lowland soil occurrences ((299/427)*100). In a similar vein, the conifer/upland coincidence is strong as it accounts for 97.98 percent of all upland soil occurrences. Both columns of coincidence percentages must be considered because a single high percent might be merely the result of the other category occurring just about everywhere.

What a bunch of droning gibberish. Maybe you better read that last paragraph again (and again). It's important because it is the basis of spatial statistics' concept of "correlation"—the direct relationship between two variables. For the non-techy types seeking just "the big picture," a coincidence table provides insight into map category relationships. A search of the table for unusually high percent overlaps of map categories uncovers strong positive relationships. Relatively low percent overlaps indicate negative relationships.

The 1 percent and 2 percent overlaps for deciduous/upland categories suggests the trees are avoiding these soils. I wonder what spatial relationship exists for spotted owl activity and forest type? For owl activity and human activity? For convenience store locations and housing density? For incidence of respiratory disease and proximity to highways?

Further Considerations

There are still several loose ends to tie before we can wrap-up point-by-point overlay summaries. One is direct map comparison, or "change detection." For example, if you encode a series of land-use maps for an area and then subtract each successive pair of maps, the locations that changed will appear as non-zero values for each time step. In GIS-speak, you would enter

```
COMPUTE LANDUSE-T2 MINUS LANDUSE-T1 FOR
CHANGE-T2&1
```

for a map of the land-use change between Time 1 and Time 2.

If you are real tricky and think digitally, you will assign a binary progression to the land-use categories (1, 2, 4, 8, 16, etc.), because the differences will automatically identify the nature of the change. The only way you can get a 1 is 2-1; a 2 is 4-2; a 3 is 4-1; a 6 is 8-2; etc. A negative sign indicates the opposite change, and now all bases are covered. Prime numbers will also work, but they require more brain power to interpret.

Our last point-by-point operation—covering—is a weird one. This operation is truly spatial and has no traditional math counterpart. Imagine that you have prepared two acetate sheets by coloring all of the forested areas an opaque green on one sheet and all of the roads an opaque red on the other sheet. Now overlay the sheets on a light-table. If you place the forest sheet down first, the red roads will "cover" the green forests, and you will see the roads passing through the forests. If the roads map goes down first, the red lines will stop abruptly at the green forest globs.

In a GIS, however, the colors become numbers and the clear acetate is assigned a zero value. The command

        COVER FORESTS WITH ROADS

causes the computer to go to a location and assess the shish kebab of values it finds. If the kebab value for roads is zero (clear), keep the forest value underneath it. If the road value is nonzero, place that value at the location, regardless of the value underneath.

So what? What's it good for? There are a lot of advanced modeling uses; however, covering is most frequently used for masking map information. Let's say you just computed a slope map for a large area and you want to identify the slope for just your district. You would create a mask by assigning zero to your district and some wild number like 32,000 to the area outside your district. Now cover the slope map with your mask, and the slopes will show through for just your district. This should be a comfortable operation. It is just like you do on the light-table.

But so much for that comfortable feeling. Let's extend our thinking to a regionwide map overlay. Imagine that you're back inside your computer, but this time you end up sandwiched between two maps. It's a horrible place, and you are up to your ankles in numbers. You glance up and note a pattern in the numbers on the map above. It is the exact shape of your district! This time you take the spear and attach a rope, like an oversized needle and thread. You wander around threading the numbers at your feet until you have impaled all of them within the boundary of your district. Now you run to the CPU, calculate their average and assign this average value to your district. Voila! You now know the average slope for your district, provided you were sloshing around in slope values.

Because you're computerized and moving at megahertz speed, you decide to repeat the process for all the other districts denoted on the template map above you. You are sort of a digital cookie-cutter summarizing the numbers you find on one map within the boundaries identified on another map. That's the weird world of regionwide map overlay. In GIS-speak, you would enter

COMPOSITE DISTRICTS WITH SLOPE AVERAGE.

However, averages aren't the only summaries you can perform with your lace of numbers. Other summary statistics you might use include the total, maximum, minimum, median, mode, or minority value; the standard deviation, variance, or diversity of values; and the correlation, deviation, or uniqueness of a particular combination of values. See, mathematics and statistics really are the cornerstones of GIS.

For example, a map indicating the proportion of undeveloped land within each of several counties could be generated by superimposing a map of county boundaries on a land-use map and computing the ratio of undeveloped land to the total land area for each county. Or a postal-code map could be superimposed over maps of demographic data to determine the average income, average age, and dominant ethnic group within each ZIP code. Or a map of dead timber stands could be used as a template to determine the average slope, dominant aspect, average elevation, and dominant soil for each stand. If the trees tend to be dying at steep, northerly, high elevations on acidic soils, this information might help you locate areas of threatened living trees that would benefit from management action. Sort of a preventative health plan for the woods.

In article 23, point-by-point and regionwide overlaying will be extended to concepts of mapwide overlay. If all goes well, this will complete our overview of map overlay, and we can forge ahead to other interesting (?) topics.

23 | If I Hadn't Have Believed It, I Wouldn't Have Seen It

(Mapwide Overlay and the Evaluation of Spatial Equations)

❖   ❖   ❖

For better or worse, much of map analysis is left to human viewing. In many ways, the analytical power of the human mind far exceeds the methodical algorithms of a GIS. As your eye roams across a map, you immediately assess the relationships among spatial features, and your judgment translates these data into meaningful information. No bits, bytes, or buts; that's the way it is. Just as you see it.

Recently, I had an opportunity to work with an organization that acquired a major GIS software package, developed an extensive database over a period of several months, and just began using the system in decision making. From more than 100 map layers in the database, three composite maps were generated for each of the 18 topographical sheets covering the project area. The three maps were aligned on top of each other, and a fourth clear acetate sheet was attached to complete the bundle.

The 18-map bundles, in turn, were edge matched and taped along the wall of a local gymnasium. A group of decision makers strolled down the gallery of maps, stopping and flipping through each bundle as they went. A profusion of discussion ensued. Finally, with knitted brows and nodding heads, they sketched annotations onto the clear top sheet, designating areas available for logging, development, wildlife habitat, recreation, and a myriad of other land uses. The set of "solution" sheets were peeled off the top and given to the stunned GIS specialists to be traced into the GIS for final plotting.

Obviously, map overlay means different things to different people. To the decision makers it provided a "data sandwich" for their visual analysis. To the GIS specialists, it not only organizes and displays map composites, but it provides new analytical tools. It means combinatorial, computational, and statistical summaries of map coincidence. As noted in articles 21 and 22, the coincidence data to be summarized can be obtained by point-by-point or region-wide map overlay techniques.

Mapwide Overlay

With these discussions behind us, we can move on to a third way of combining maps, mapwide overlay. Recall that point-by-point overlay can be conceptualized as vertically "spearing" a shish kebab of numbers for each location on a set of registered maps. By contrast, regionwide overlay horizontally "laces" a string of numbers within an area identified on another map. Now, are you ready for this? Mapwide overlay can be thought of as "plunging" an equation through a set of registered maps. In this instance, each map is considered a variable, each location is considered a case, and each value is considered a measurement. These terms (variable, case, and measurement) hold special significance for techy types and have certain rights, privileges, and responsibilities when evaluating equations. For the rest of us, it means that the entire map area is summarized in accordance with an equation.

Map Comparison and Similarity

For example, mapwide overlay can be used to compare two maps. Traditional statistics provides several techniques for assessing the similarity among sets of numbers. The GIS provides the organization of the number sets—cells in a raster system and polygonal progeny in a vector system. A simple t- or F-test uses the means and variances of two sample populations to determine if one can statistically say, "They came from the same mold." Suppose two sample populations of soil lead concentration cover the same map area for two different time periods. Did lead concentration significantly change? Mapwide comparison statistics provide considerable insight.

Another comparison technique is similarity. Suppose you have a stack of maps for the world depicting the gross national product, population density, animal protein consumption, and other variables describing human living conditions. Which areas are similar and which areas are dissimilar? In the early 1980s, I had a chance to put this approach into action. The concept of *regionalism* reached its peak, and the World Bank was interested in ways it could partition the world into similar groupings. A program was developed allowing the user to identify a location (say northeastern Nigeria) and generate a map of similarity for all other locations of the world. The similarity map contained values from 0 (totally dissimilar) to 100 (identical).

A remote sensing specialist would say, "So what? No big deal." It is a standard multivariate classification procedure. Spear the characteristics of the area of interest (referred to as a *feature vector*) and compare this response pattern to those of every other location in the map area. The remote sensing specialist is right; it is no big deal. Scale adjustments are all that are needed to normalize map response ranges. The rest is standard multivariate analysis. However, to some it is mind wrenching because we normally do not mix map analysis and multivariate classification in the same breath. But that's before the digital map took us beyond mapping.

Let's try another application perspective. A natural resource manager might have a set of maps depicting slope, aspect, soil type, depth to bedrock, etc. Which areas are similar and which areas are dissimilar? The procedure is like the one previously mentioned. In this instance, however, clustering techniques are used to group locations of similar characteristics. Techy terms of *intracluster* and *intercluster distances in multivariate space* report the similarities. To the manager, the map shows ecological potential, a radically different way to carve the landscape.

Chances are, the current carving of the forest into timber management units was derived by aerial photograph interpretation. Some of the parcels are the visible result of cut-out/get-out logging and forest fires that failed to respect ecological partitions. Comparison of the current management parcels to the ecological groupings might bring management actions more into line with Mother Nature. The alternative is to manage the woods into perpetuity based on how the landscape exposed itself to aerial film several years back.

Mapping Predictive Equations

In addition to comparison and similarity indices, predictive equations can be evaluated. For example, consider an old application done in the late 1970s. A timber company in the Pacific Northwest was concerned with predicting timber felling *breakage*. You see, when you cut a tree, there is a chance it will crack when it hits the ground. If it cracks, the sawmill will produce little chunks of wood instead of large, valuable boards. This could cost you millions. Where should you send your best teams with specialized equipment to minimize breakage?

Common sense tells you if there are big old rotten trees on steep slopes, expect kindling at the mill. A regression equation specifies this more rigorously as

$$Y = -2.490 + 1.670X1 + 0.424X2 - 0.007X3 - 1.120X4 - 5.090X5$$

where Y = predicted timber felling breakage and $X1$ = percent slope, $X2$ = tree diameter, $X3$ = tree height, $X4$ = tree volume, and $X5$ = percent defect.

Now you go to the woods and collect data at several sample plots, then calculate the average for each variable. You substitute your averages into the equation and solve. There, that's it—predicted timber felling breakage for the proposed harvest unit. If it is high, send in the special timber beasts. If it is low, send them elsewhere.

But is it really that simple? What if there are big trees to the north and little trees to the south? Your analytical procedure assumes that there must be medium-sized (average) trees everywhere. And what if it is steep to the north and fairly flat to the south? It must be moderately sloped (average) everywhere is the assumption. This reasoning leads to medium-sized trees on moderate slopes everywhere. Right? But hold it, let's be spatially specific, there are big trees on steep slopes to the north. This is a real board-and-profit-busting condition. Your field data are trying to tell you this, yet your nonspatial analysis blurs the information into typical responses assumed the same everywhere.

As depicted in figure 23.1, a spatial analysis first creates continuous surface maps from the field data (see also article 10). These mapped variables, in turn, are multiplied by their respective regression coefficients and then summed. The result is the spatial evaluation of the regression equation in the form of a map of timber felling breakage. Not a single value assumed everywhere, but a map showing areas of higher and lower breakage. That's a lot more guidance from the same set of field data.

Spatial Autocorrelations

In fact, further investigation showed the overall average of the map predictions was very different from the nonspatial prediction. Why? It was due to our final concept, spatial autocorrelation. The underlying

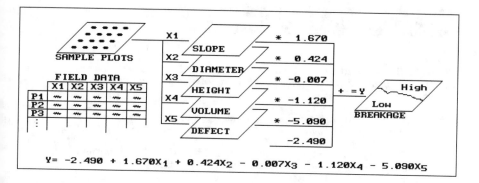

Fig. 23.1. Spatially evaluating prediction equations.

assumption of the nonspatial analysis is that all of the variables of an equation are independent. This is a poor assumption, particularly in natural resource applications. People, birds, bears, and even trees tend to spatially select their surroundings. It is rare to find any resource randomly distributed in space. In this example, there is a compelling and easily explainable reason for big trees on steep slopes. During the 1930s, the area was logged, and who in their right mind would hassle with the trees on the steep slopes. The loggers just moved to the next valley, leaving the big trees on the steeper slopes—an obvious relationship, or more precisely termed, spatial autocorrelation. This condition affects many of our quantitative models based on geographic data.

So where does all this lead? To at least an appreciation that map overlay is more than just a data sandwich. The ability to render graphic output from a geographic search is just the beginning of what you can do with the map overlay procedures embedded in your GIS.

Recommended Reading

Books

Burrough, P.A. "Methods of Data Analysis and Spatial Modeling." *Principles of Geographical Information Systems for Land Resources Assessment.* Oxford, UK: Oxford University Press, 1987.

Epstein, E., and P. Brown. "Introduction to Multipurpose Land Information Systems." *1991-92 International GIS Sourcebook,* 371-76. Fort Collins, CO: GIS World, 1991.

Maguire, D., and J. Dangermond. "The Functionality of GIS." In *Geographical Information Systems: Principles and Applications,* ed. D.J. Maguire, M.F. Goodchild, and D.W. Rhind, Vol. 1, 319-35. Essex, UK: Longman, 1991.

McHarg, I. *Design with Nature.* Garden City, NJ: Doubleday, 1969.

Openshaw, S. "Developing Appropriate Spatial Analysis Methods for GIS." In *Geographical Information Systems: Principles and Applications,* ed. D.J. Maguire, M.F. Goodchild, and D.W. Rhind, Vol. 1, 389-402. Essex, UK: Longman, 1991.

Star, J., and J. Estes. "Manipulation and Analysis." Chapt. 8 in *Geographic Information Systems: An Introduction,* Englewood Cliffs, NJ: Prentice Hall, 1990.

Thomas, R., and R. Huggett *Modelling in Geography: A Mathematical Approach.* London: Harper & Row, 1980.

Tomlin, C.D. "Characterizing Individual Locations" and "Characterizing Locations Within Zones." Chapts. 4 and 6 in *Geographic Information Systems and Cartographic Modeling.* Englewood Cliffs, NJ: Prentice Hall, 1990.

Journal Articles

Anselin, L., and D. Griffith "Do Spatial Effects Really Matter in Regression Analysis?" *Papers of the Regional Science Association* 65: 11-34 (1988).

Goodchild, M.F. "A Spatial Analytical Perspective on Geographical Information Systems." *International Journal of Geographical Information Systems* 1: 327-34 (1988).

Kennedy, S. "A Geographical Regression Model for Medical Statistics." *Social Science and Medicine* 26: 119-29 (1988).

Meyers, D.E. "Multivariate Geostatistics for Environmental Monitoring." *Sciences de la Terra* 27: 411-27 (1988).

Steinitz, C.F., et al. "Hand Drawn Overlays: Their History and Prospective Uses." *Landscape Architecture* 66: 444-55 (1976).

Woodcock, C., et al. "Comments on Selecting a Geographic Information System for Environmental Management." *Environmental Management* 14: 307-15 (1990).

TOPIC 8

SCOPING GIS: WHAT TO CONSIDER

You go duck hunting where the ducks are.

–Lyndon B. Johnson

GIS technology is a radical departure from traditional map processing; therefore, assessing its potential within an organization needs to go beyond traditional cost-benefit analysis. This section describes the major organizational, social, and personal ramifications of implementing GIS. Both formal and informal GIS training considerations are discussed.

24 Both Dreams and Nightmares Are Born of Frustration

(The Limitations of Traditional Cost-Benefit Analysis)

❖   ❖   ❖

The dream is that GIS can do anything; the reality is that it isn't easy. With increasing fervor, technologists and users alike define and redefine the "unlimited" potential of GIS technology. These dreams are, at least in part, an expression of our hopes, as well as our science. When considering if GIS is for you, often your biggest challenge is to carefully separate what you hear into two distinct piles, the quixotic dream and the pragmatic reality.

The first step in this process is establishing where you are coming from. GIS means different things to different people. At least four distinct perspectives flavor both our expectations and our realities: economic, organizational, visionary, and emotional. The economic perspective is usually based on labor and time-saving considerations. Standard cost-benefit analyses are particularly appropriate in distilling dreams from reality. A careful audit of your organization's current mapping and spatial data handling procedures establishes a reference to estimate the savings in moving "from pen to plotter and from file drawer to keyboard." If the savings are greater than the expenditures, you are economically irrational (foolish) if you don't implement GIS immediately.

There, that's easy. There is nothing to it. Just call in the accountants, and they will identify the numbers to plug into the cost-benefit equation. The reality is that even a strictly economic perspective is not that easy. The comfortable feeling of quantifying the evaluation process is quickly lost to the pliable nature of the "yardsticks" used to measure the costs and benefits.

The time span used in the analysis is critical. If the span is too short, the stream of benefits is artificially truncated. The high front-end costs, combined with the confusion and frustration of implementing a new system, will far outweigh the benefits. It's like a bare-knuckle battle between Sylvester Stallone and a tiger cub. If it is delayed a few years, the outcome will likely be different. If you had used a two-week cost recovery period for word processing, would you have ever dropped your pencil?

So what time period should be used? That's a judgment call. Like lying with statistics, you can choose the time period that ensures the answer you want. In general, a long-term position favors the GIS adoption.

Hardware/Software Costs

Just as important is how you identify and quantify the variables of the cost-benefit equation. Four cost considerations quickly surface: hardware/software, database development, training, and application models. The hardware figures are the easiest to quantify through a litany of parameters, including megahertz, gigabytes, RAM, SIMMS, MIPS (DIPS, DRIPS, and SLIPS). Software specifications are a bit more difficult, yet factors such as point-in-polygon, buffering, coordinate accuracy, and transfer formats can be used.

Although relatively easy to quantify, these figures are fleeting and set you up for a bad case of "buyer's remorse." When you finally push through your procurement and take first delivery, your system is out of date. It's like a pocket calculator. Within a couple of months, the same expenditure gets you five more keys at half the price. The difficulty in nailing down the hardware/software cost component isn't the definitions; it's keeping your footing in the quicksand of technology. Like shooting ducks, you better have a good lead on your target. For large, bureaucratic organizations, it may be prudent to set a budgetary figure for the "best-available" technology and postpone the specifications to the moment of purchase. That may seem preposterous, but it may be more realistic.

Database Costs

Database development is not only a larger expense than hardware and software, but its cost is even more tricky and slippery to estimate. Rarely does a simple inventory of your current map and file cabinets multiplied times an estimate of encoding costs produce an acceptable cost figure. The differences between the digital and paper map make it too tricky for such a mechanical approach. It's prudent to launch an information needs assessment to determine database contents, structure, and costs.

Even if you get a good handle on the database you must develop, you're not out of the woods yet. How you obtain these data is slippery turf. Manual encoding, scanning, or purchasing are your basic options. Not long ago, in-house, manual encoding was your only option. More recently, the scales have been tipping toward scanning and purchasing because a room full of digitizer folks is a major cost and a distraction from normal business activities. Also, many of the maps you might encode have time bombs ticking in them. For example, if you encode (in-house or contract) a soils map, it will become invalid once the U.S. Soil Conservation Service's "authoritative" version is released. It's back to shooting ducks; you had better get your data requirements in line and lead them, or you will just be pumping pellets into the air.

GIS Training

The costs of training your people to use GIS can easily outstrip the combined costs of hardware/software and database development. Early GIS successes were often more a function of the zealots using it than the technology itself. Like the Little Engine That Could, GIS could do a lot. The pitfalls that accompany any new technology are overcome by innovative "work arounds" of committed users. In a wholesale adoption, however, the user community is expanded to "I don't think I can" and "I am damned if I will" outlooks.

One reaction to this reality is to form a GIS division. On the surface, it's a plausible alternative. All you have to do is train a small cadre of experts. There, that's both efficient and effective. But it rarely works for two reasons.

First, the GIS product produced is just that, a GIS product and not the direct expression of the final user. In the late 1970s I had an opportunity to observe a large timber company's centralized GIS implementation. Most of the field personnel merely dismissed the "computer jerk's" forest management maps handed to them through the glass windows of the computer center. "What do they know about the @#*^! forest anyway?" was the rallying cry. If the maps were used at all, they became the center of attention for the short period it took to locate that one forest stand in the middle of a lake. This meant wasted effort on both the GIS and user sides, a situation that could be helped with sufficient investment in training.

If training costs are identified at all, they are usually associated with vocational instruction on system operations. But GIS is a challenging new way of thinking, as well as a new sequence of buttons to push. The mechanics of translating what you currently do with maps into a GIS is straightforward. In fact, colorful icons and mouse clicking can make it almost fun. However, most potential GIS applications within an organization are yet to be discovered.

Application Models

The development of application models is the second reason for failure of a centralized cost-benefit approach. How the new technology leads to new ways of doing things is the least-understood cost (and benefit) of GIS technology. It's like your son or daughter dumping the tin of tinkertoys™ on the floor. The mechanics of how the pieces fit together are fairly simple. What ought to be built with the individual pieces is the difficult part. The tinkertoy makers (GIS experts) can supply some ideas, but they certainly do not cover all possibilities. Vocational training develops an awareness of the GIS "owner's manual" description of the pieces and parts, but beware "some assembly is required" before you are up and running.

The creative assembly is entirely up to your people. If you ignore or skimp on training and application model development, you will incur opportunity costs at the minimum. More likely, you will generate a backlash of confusion and apprehension that quickly outweighs the benefits. A couple of strategically placed anti-GIS terrorists will wreak havoc with even your best laid plans.

A strict economic perspective is the first step in scoping GIS technology. Identification, and ultimately quantification, of the costs and benefits sets the stage. However, organizational, visionary, and emotional perspectives are needed to complete the picture—whether a dream or a nightmare.

25 GIS Is Never Having to Say You're Sorry

(Human and Organizational Considerations)

❖   ❖   ❖

In the previous article I discussed the broad considerations of implementing a GIS with an economic cost-benefit analysis. This is where most organizations take their first step of what seems to be a 1,000-mile journey to GIS implementation. At first glance, the seductive appearance of a rigorous, quantitative analysis is quickly lost to the pliable nature of the "yardsticks" used to measure the costs and benefits. At best, a cost-benefit analysis sets the stage for further investigation into the full impact of implementing a GIS. Even the most favorable cost-benefit ratio should be further scrutinized in terms of the organizational and human impacts of GIS. Whether real or imagined, the perceived threats of GIS technology form the actual mine field you must traverse.

Organizational structure (both formal and informal) is an important concern, because it is the direct expression of the *corporate character*, the most basic element of any organization. If extensive individual latitude and autonomy best describe an organization's current character, GIS will likely have a rocky road to implementation. Within this environment, data often are viewed as the medium of exchange for power brokers at all levels. Simply stated, "If you must pass through me to get to important data in my map cabinets and file drawers, then I am as important as the data I keep." However, if GIS places my data in some central repository accessible to all by a single mouse click, my corporate worth has been severely devalued. The result, as viewed by some, is an electronic end run around the current data gatekeepers and a direct assault on the existing organizational structure. It may be a benefit to the organization to have a corporate database, but to many it represents a personal loss of influence. If your implementation plan ignores this reality, you'll be sorry.

Another concern that may run amok with the corporate character is the imposition of data standards. In many organizations, mapping standards are nonexistent or merely address geographic registration and data exchange formats. But this is just the tip of the chilling iceberg of standards. The ability to export a map from one GIS package and swallow it in another is basic and rapidly becoming a nonissue. Likewise, the ability to convert projections, as well as rectify and register maps, is commonplace (although not necessarily easy). The confusion and frustration isn't in the locational (where) set of standards, but in the informational (what) set. Article 33 develops this point in greater detail.

The Corporate Database

A corporate database consists of three levels of maps based on their degree of abstraction: base, derived, and interpreted. Base maps are usually physical data we collect such as roads, water, and ownership boundaries. They have minimal abstraction from reality and, as much as possible, represent a scale model with all of the detail of a flattened model train set. Definitions and procedures for mapping most of these data are in place, but not all.

Consider a map of cover type. Is "forest"/"nonforest" a sufficient standard? Or should the "forest" class be further divided into "conifer" and "deciduous"? And should the "conifer," in turn, be subdivided into "pine," "fir," and, "hemlock"? What about age and stocking classes? Should you identify a lone pine tree in the middle of a meadow as a "conifer stand"? Two, three, four, five trees—what does it take to form a forest stand? Ask a forester, ecologist, and recreation scientist, and you'll get at least three different responses. Or maybe four or five different responses depending on how they decipher different applications. You'll be sorry if you don't tackle these questions before you implement a GIS.

For example, a wildfire had the audacity to burn across the boundary of two national forests. Maps of cover type were encoded for both forests, but they couldn't be edge matched. One forest had six classes of age and stocking for Douglas fir, the other had eight. The GIS was able to account for locational adjustments during encoding but not for the differences in informational content. A common classification standard for cover type had to be established and encoded. The struggle for whose classification scheme was the best eclipsed the mundane tasks of reconstructing and encoding a compatible cover type map. The challenges to human and organizational interests run much deeper than those encountered at the digitizing tablet.

Vested interests in the definitions of map categories go beyond base data. Derived maps, such as slope, visual exposure, and proximity to roads, are physical things. However, the data are too difficult to collect, so we use the computer to calculate them. Even something as simple as slope calculation has several algorithms, each with its pros and cons. For something as complex as visual exposure, there is a quagmire of assumptions, approaches, and procedures. Which will you entrench in your system? Rest assured that the choice won't be by consensus nor the dissenting voices reserved.

Even more volatile are the assumptions embedded in interpreted maps. These data are the most abstract because they are conceptual renderings of expert opinion. Taunts of "my elk habitat model is better than yours" reverberate through the halls whenever two wildlife ecologists are cornered in the same room. It is naive to assume that an elk model will edge match across two forests, much less an entire region, and certainly not across the paradigm chasm of two experts.

How to Be Sorry

So whose derived and interpreted maps capture the standards in the corporate database? The question of standards runs a lot deeper than just geographic registration and encoding effort. It involves organizational and individual perceptions, reputations, and vested interests. You'll be sorry if your implementation plan ignores these elements. Sure, they will get sorted out later—after you and the GIS fail.

A GIS implementation strategy has to go beyond simply scoping system design to nurturing a receptive environment. This passes the baton from the system engineers and GIS specialists to the sociologists and human relation professionals. As I continually remind you (possibly to the point of being shrill), GIS is not just automating what you do but changing how you do things. Sensitivity to the full impact of these changes, human as well as procedural, is paramount.

Table 25.1 outlines some of the threats and responses that need to be addressed. The table is designed to stimulate discussion in a workshop setting, but I hope it will trip some thoughts in your mind. As you look over the outline, try some free associations with the points. Conjure up some of your own threats and possible coping responses. It is a lot of fun at the workshops and sparks a broader perspective on GIS implementation. At minimum, the exercise should encourage you to go beyond a focus on the mechanics of GIS technology to its institutional and human implications. If you don't, you'll be sorry.

Table 25.1. Coping with GIS threats.

GIS Threats

Institutional Threats

- Organizational –"There's only one problem having all this sophisticated equipment, we don't have anyone sophisticated enough to use it."

- Status Quo – "If it ain't broke, don't fix it."

- Overload – "Torture numbers and they will tell you anything."

- Stifling – "Imagination is more important than information."

- Awareness –"Technobabble, that seemingly endless drone masking what otherwise would be a clear understanding of a new technology's concepts and use."

Personal Threats

- Intimidation – "It's like new math, I'm just too old."

- Power – "Experience used to be worth something, now you just dazzle them with color."

- Dependence – "Middle management is an endangered species ... they are information brokers hooked to the computer jerk down the hall."

Coping with GIS Threats

Choices – Fight It • Ignore It • Face It

Institutional Response

- Grass Roots Support – "They don't know what they are doing."

- Understanding – "You know, this GIS stuff isn't so bad after all."

- Proof of Concept – "Oh, now I see, I could use something like that."

- Commitment – "What do you mean, learn it in my spare time."

- Tough Love – "Like it or not, unless you have retired, your job has evolved for the better."

Personal Response

- Lingo – "Sticks and stones may break my bones, but arcane terminology will never hurt me."

- Continuing Education – "The era of the four-year smart pill is over."

- Leadership – "If the boss can handle this stuff, then I guess anybody can."

- Long Haul – "Like other new technologies, GIS is something that is best understood backward but must be learned forward."

---

26 A Tailored Plan and Curriculum Cure GIS Training Woes

(The Importance of Effective Education and Training)

❖   ❖   ❖

Why waste time learning when ignorance
is instantaneous?

–Calvin and Hobbes

The two previous articles encouraged you to go beyond a focus on the mechanics of GIS to its institutional and human implications. Like most new technologies, the technical aspects of GIS are the easiest parts. The nontechnical implications and impacts ultimately determine success or failure. These have little to do with bits, bytes, buffers, or even bucks. It's a corporate "warm, fuzzy feeling" about GIS.

So how does one ensure such acceptance? Let's start with the easier question, "How does one ensure failure?" That's simple. Just deliver crates of computers and masses of shrink-wrapped manuals. Within hours, anti-GIS terrorists will have torched all the managerial offices and be moving toward the board room. Defensive positions such as, "We're doing it for you" and "Try it, you'll like it" will crumble like papier-mâché bunkers. General rebellion and anarchy will sap any remaining vestiges of the corporate "good idea." An emotional and intellectual wasteland will lie at the feet of the sterile gray boxes and brightly colored CRTs.

Like the vision of "Christmas Future" in Dickens's novel, *A Christmas Carol*, there is an alternative. It's a commitment to education. Without it, ignorance prevails, confusion is rampant, and negative rumors abound. However, at least as much effort is involved in planning and implementing an effective educational program as in scoping hardware and software requirements. Matching instructional approach with required skills is similar to matching appropriate platform and functionality to an organization's information needs.

Effective Training Considerations

But why not leave training to the universities? It's their job isn't it? Two major points come to mind. First, most universities have tightly defined programs for traditional degree-seeking students. It is hard enough to get a GIS course into a recognized major, let alone canonized as a new program of study. However, even if the academic tanker was turned overnight, you can't wait until the GIS matriculates rise to top decision-making positions. Technology moves faster than scholarly debate or openings in your organization. You're left with nontraditional students—your current employees with all of the warts and scars left from their last brush with the "four-year" smart pill. A snarly bunch, but they're the key to the success or failure of GIS in your organization.

The first consideration in GIS training is a recognition that there are both formal and informal processes. Attention to the informal process must be made throughout the GIS implementation by nurturing "in-house zealots." In the early stages, these individuals not only provide over-the-shoulder instruction, but legitimize the technology. They enthusiastically demonstrate that "one of us" can use the damn thing. Your challenge is to create situations that quickly identify these individuals. A good strategy is undertaking a couple of pilot projects early in the GIS scoping process. However, be certain that the volunteers can focus their full attention on the project, receive ample support, and are given complete freedom in their approach. A good rule of thumb is that if you can't afford this involvement, the chances are you can't afford the technology. Another good rule is that one in-house GIS zealot is worth a dozen out-house specialists.

Training Approaches

The zealots provide leadership and credibility but are inefficient in conveying basic procedures and concepts to the masses. This is where formal training comes in. Three instructional approaches are involved: awareness, vocational, and educational.

General awareness instruction provides a nontechnical overview of GIS technology's capabilities and limitations. Such instruction counters GIS ignorance but stops far short of a working knowledge of the field. It is imperative that all personnel participate in this training phase—from clerical to technical to professional to managerial. Ideally, the presentation is made in mixed audiences and discussion is encouraged. It is not so much a tutelage as it is a forum. Sure, some basic concepts and terminology slip in, but the gathering introduces GIS and sets the stage for its implementation. The mixed audience provides recognition of in-house zealots and reinforces management's commitment. The absence of GIS experts or top management is dysfunctional because it makes the meeting merely perfunctory.

Vocational instruction, on the one hand, develops operating skills in the procedures and practices of a specific system. It is designed to show you how to use the system.

Educational instruction, on the other hand, develops spatial reasoning skills through understanding of basic concepts and theory. It is designed to show you why you might want to use the system.

Think of it this way—awareness instruction is similar to a newspaper article; vocational instruction is similar to a manual; and educational instruction is similar to a textbook. Each approach or item is directed to a different audience, presents different material, and produces different products. An inappropriate match bores or overwhelms the audience, in either case rendering your training a waste of time and money.

But enough of the academic hyperbole; what's the reality? The reality is that most implementation plans focus on vocational training alone (if at all). Remember when your organization implemented word processing? Did you receive instruction, or were you left to your own devices? How did you do? How would you do it differently? GIS is like word processing, only different. Compared to manual techniques both are faster and easier and can generate more piles of paper than wall space and surface area of furniture combined. But is GIS better? Probably not until you use a new technology in new ways. That's the big difference; GIS presents an entirely new way of doing things. In addition to the mechanics of how to work the thing, new analytic concepts and spatial reasoning skills must be developed.

Striking a Balance

The balance between vocational and educational approaches depends on which skills are addressed. Four distinct GIS skill levels can be identified.

- *Database development technician* encodes and maintains the spatial and attribute databases;

- *Data center manager* coordinates data integration, information flow and maintains the system;

- *Application specialist* facilitates the development of application-specific models; and

- *General user* uses GIS in both routine activities and decision making.

Note that the first three roles support the users and their ability to do the job. That's not a moot point. Without frequent reality checks, any new technology can take on a life of its own. Also, the listing is ordered in terms of the GIS technical knowledge required (from most to least). In general, more technical/vocational training is needed at the top of the list. The balance shifts to more conceptual/educational training at the bottom. So what? At a minimum, you're put on alert that one "comprehensive" short course may not be sufficient for all GIS skill levels.

So what's the appropriate balance? As usual, it depends, and mostly on your information needs (i.e., the user). If routine mapping and database management demands dominant system use, then training is best focused on the top of the list. Training in the proper care and feeding of the database is paramount. The specialist and user interests revolve around graphic user interfaces that make access and retrieval a "piece of cake." The bulk of automated mapping/facilities management (AM/FM) applications fall into this group.

However, if your GIS needs to lean more toward decision support systems (DSS), then your training requirements are significantly altered and move toward a more conceptual/educational focus. In this environment, GIS is less a tool to extend the hand than a medium to extend the mind. Creative uses, such as spatial modeling, require both a proficiency in system operation and a thorough understanding of system functionality. Effective dialogue between the application specialist and the general user is rooted in a common understanding of GIS capabilities and limitations in expressing spatial relationships. Even memorization of the user's manual won't cover these bases. An organizational commitment to education will.

Some say, "Ignorance is instantaneous and a lot cheaper," but is it really? This discussion should have dispelled the notion of GIS training through "immaculate conception." If your personnel struggled with word processing, choked on spreadsheets, and gagged on database management, expect to be in intensive care with a massive head wound with GIS. Preventive medicine, in the form of a tailored training plan and curriculum, is advised. At least as much thought (and ultimately, direct investment) should go into training as in scoping hardware/software and database requirements.

Recommended Reading

Books

Aronoff, S. "Implementing a GIS" and "Conclusion." Chapts. 8 and 9 in *Geographic Information Systems: A Management Perspective*. Ottawa, Canada: WDL Publications, 1989.

Burrough, P.A. "Choosing a Geographical Information System." Chapt. 9 in *Principles of Geographical Information Systems for Land Resources Assessment*. Oxford, UK: Oxford University Press, 1987.

Clarke, A.L. "GIS Specification, Evaluation and Implementation." In *Geographical Information Systems: Principles and Applications*, ed. D.J. Maguire, M.F. Goodchild, and D.W. Rhind, Vol. 1, 477-88. Essex, UK: Longman, 1991.

Dangermond, J. "A Review of Digital Data Commonly Available and Some Practical Problems of Entering Them into a GIS." In *Fundamentals of Geographic Information Systems: A Compendium*, ed. W. Ripple, 41-58. Bethesda, MD: American Society of Photogrammetry and Remote Sensing, 1989.

Epstein, E.F. "Legal Aspects of GIS." In *Geographical Information Systems: Principles and Applications*, ed. D.J. Maguire, M.F. Goodchild, and D.W. Rhind, Vol. 1: 489-502. Essex, UK: Longman, 1991.

Montgomery, G.E. "Guide to Project Implementation." *The GIS Sourcebook*, 133-42. Fort Collins, CO: GIS World, 1989.

Openshaw, S. "Why Are We Still Waiting for Spatial Analysis Functionality?" *The 1990 GIS Sourcebook*, 13-16. Fort Collins, CO: GIS World, 1990.

Openshaw, S., and C. Brunsdon. "An Introduction to Spatial Analysis in GIS." *1991-92 International GIS Sourcebook*, 401-4. Fort Collins, CO: GIS World, 1991.

Star, J., and J. Estes. "Practical Matters." Chapt. 11 in *Geographic Information Systems: An Introduction*, Englewood Cliffs, NJ: Prentice Hall, 1990.

Journal Articles

Berry, J.K. "Learning Computer-Assisted Map Analysis." *Journal of Forestry*: 39-43 (October 1986).

Goodchild, M., and B. Rizzo. "Performance Evaluation and Work-Load Estimation for Geographic Information Systems." *International Journal of Geographical Information Systems* 1: 67 (1987).

Guptill, S.S. "Evaluating Geographic Information Systems Technology." *Photogrammetric Engineering and Remote Sensing* 55(11): 1583-88 (1989).

Lai, P.C. "Issues Concerning the Technology Transfer of Geographic Information Systems." *Environmental Management* 15: 595-601 (1991).

TOPIC 9

SLOPE, DISTANCE, AND CONNECTIVITY: THEIR ALGORITHMS

You don't know what you don't know.

–Anonymous

At first encounter, many of the advanced GIS analytical operations are intimidating. However, a basic understanding of the computer's procedures is needed to assess the potential and limitations of the new tools. This section describes various approaches used in computing slope, effective distance, optimal paths, and visual connectivity.

27 There's More Than One Way to Figure Slope

(Calculating Slope and Its Impact on Applications)

❖　　❖　　❖

You knowest me not.

–Romeo and Juliet

Romeo speaks these famous lines as he attempts to befriend his family's enemy, Tybalt. However, as the scene unfolds, Romeo wields his sword and runs him through. Maybe Tybalt knowest him too well? No, it's just plain old ignorance. Romeo knowest his secret marriage to Juliet maketh him Tybalt's brother, but Tybalt knowest not such portentous information.

What's all this got to do with GIS? A simple lesson: knowest thy GIS operations, or thee shalt be gored. Take the simple concept of slope. We all remember Ms. Deal's explanation of "rise over run." If the rise is 10 feet when you walk (or run) 100 feet, then the slope is 10/100 * 100 percent = 10 percent slope. The concept is easy. So is the practice. Boy scouts and civil engineers should recall sighting through their clinometers and dumpy levels to measure slope.

Questions and Answers

But what's a slope *map*? Isn't it just the same concept and practice jammed into a GIS? Or is it something you may knowest not? In a nutshell, a slope map is a continuous distribution of slope (rise over run) throughout an area. Or, stated another way, it is the trigonometric tangent taken in all directions. Or it is the three-dimensional derivative of a digital terrain model. You might be thinking, "Bah! This stuff is too techy. I'll just ignore it and stick with Ms. Deal's comfortable concept."

No, try sticking it out. An overview of this subject was made in article 8. But for now let's get down to the nitty-gritty, how you derive a slope map. In most offices it is an ocular-manual process.

You look at a map and, if the contour lines are close together, you mark the area as "steep." If they are far apart, you circle it as "gentle." Anything you didn't mark must be "moderate." As a more exacting alternative, you could count the number of contour lines per unit of distance and calculate the percent slope, but that's a lot of work. You might construct a rulerlike guide but using it is still inhumane.

More importantly, the procedure assesses slope along a single transect. As you stand on the side of a hill, isn't there a bunch of different slope transects down, up, and across from you? A water drop recognizes the steepest slope. A weary hiker responses the minimum slope. A developer usually assesses the overall slope. In the real world, we need several different slope algorithms for different application requirements.

An interesting physical slope procedure involves creating a black-and-white negative of a map's contour lines. In printing the negative, the paper is rapidly "wiggled" about a pivot at its center. The areas with closely spaced contours (steep) do not expose the paper to as much light as the areas where the contours are far apart (gentle). The result is a gray-scale map progressing from dark areas with gentle slopes to light areas with steep slopes. Now we have a real slope map— a robust range of slope conditions and a procedure considering all of the slopes about each location throughout an area.

In a somewhat analogous fashion, a series of grid cells can be over-laid on a map of contour lines. The number of lines within each cell is counted, the more lines the steeper the slope. This approach was used in early vector systems but requires considerable computation and is sensitive to the spacing and positioning of the analysis grid.

Most slope processing is done with data formatted as a digital elevation model (DEM). The DEM uses a raster format of regularly spaced elevation data. Figure 27.1 shows a small portion of a typical data set, with each cell containing a value approximating its overall elevation. As there is only one value in each cell, we can't compute a slope for an individual cell without considering the values surrounding it. In the highlighted 3-x-3 window, there are eight individual slopes, as shown in the calculations accompanying figure 27.1. The steepest slope of 52 percent is formed with the center and the cells to its northwest and west. The minimum slope is 11 percent in the southwest direction.

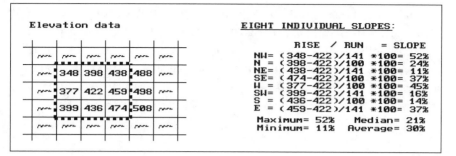

Fig. 27.1. Example elevation data.

But what about the general slope throughout the entire 3-x-3 analysis window? One estimate is 30 percent, the average of the eight individual slopes. Another general characterization could be 31 percent, the median of slopes. To get an appreciation of this processing, shift the window one column to the right and run through the calculations using the value 459 as the center. Now imagine doing that a million times with an impatient and irate human pounding at your peripherals. Show some compassion for your computer—the basis of a meaningful GIS relationship.

Alternative Approaches

An alternative format to DEM is triangulated irregular network (TIN) using irregularly spaced elevation points to form a network of "triangular facets." Instead of regular DEM spacing, areas of rapidly changing terrain receive more points than relatively flat areas. Any three elevation points, when connected, form a triangle floating in three-dimensional space. Simple (ha!) trigonometry is used to compute the slope of the tilted triangular facets, which are stored as a set of adjoining polygons in conventional vector format. TIN algorithms vary widely in how they develop the irregular network of points, so their slope maps often are considerably different.

A related raster approach treats the octants about the center position in the 3-x-3 window as eight triangular facets. The slopes of the individual facets are computed and generalized for an overall slope value. To bring things closer to home, the elevations for the cardinal points along the center cell's sides can be interpolated from the surrounding DEM values. This refinement incorporates the lateral interactions among the elevation points and often produces different results than those directly using the DEM values.

Now let's stretch our thinking a bit more. Imagine that the nine elevation values become balls floating above their respective locations, as shown in figure 27.2. Mentally insert a plane and shift it about until it is positioned to minimize the overall distances to the balls. The result is a "best-fitted" plane summarizing the overall slope in the 3-x-3 window.

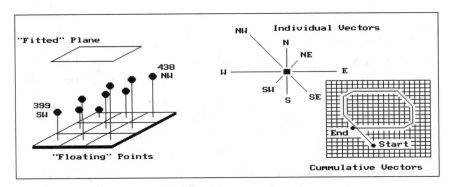

Fig. 27.2. "Best-fitted" plane using vector algebra.

The techy types will recognize this process as similar to fitting a regression line to a set of data points in two-dimensional space. In this case, it is a plane in three-dimensional space. It is an intimidating set of equations with a lot of Greek letters and subscripts to "minimize the sum of the squared deviations" from the plane to the points. Solid geometry calculations, based on the plane's "direction cosines," determine the plane's slope (and aspect).

This approach can yield radically different slope estimates to the others discussed. Often the results are closer to your visceral feeling of slope as a stand on the side of a hill; often they are not. Consider a mountain peak or a terrain depression. Geometry characterizes these features as a cone with eight steeply sloped sides; the result is very steep. The fitted-plane approach considers these features as a balanced transition in slopes; the result is perfectly flat. Which way do you view it?

Another procedure for fitting a plane to the elevation data uses vector algebra, as illustrated in the right portion of figure 27.2. In concept, the mathematics draw each of the eight slopes as a line in the proper direction and relative length of the slope value (individual vectors). Now comes the fun part. Starting with the northwest line, successively connect the lines as shown in figure 27.2 (cumulative vectors). The civil engineer will recognize this procedure as similar to the latitude and departure sums in closing a survey transect. The physicist sees it as the vector sums of atomic movements in a cloud chamber. We think of it as the serpentine path we walked in downtown Manhattan.

If the vector sum happens to close on the starting point, the slope is 0 percent. A gap between the starting and end points is called the *resultant vector*. The length of the resultant vector is the slope of a plane fitting the floating data points, which in turn is interpreted as the general slope of the terrain. In the example, the resultant slope is 36 percent oriented toward the northwest.

Considering our calculations, what's the slope, 52, 10, 30, 31, or 36? They're all technically correct. The point is, it depends on your understanding of the application's requirement and the algorithm's sensitivity. A compatible marriage between application and algorithm must be struck. Your last words might be, "With this kiss, I die" if you knowest not your GIS operations and blindly apply any old GIS result from clicking your mouse on the slope icon.

28 Distance Is Simple and Straightforward

(Calculating Simple Distance as a Propagating Wavefront)

An overview of the evolving concept of distance was made in articles 4 through 6. My objective for revisiting the subject is to focus on the real stuff—how a GIS derives a distance map. The basic concept of distance measurement involves two things, a *unit* and a *procedure*. The tic marks on your ruler establish the unit. If you are an old woodcutter like me, a quarter of an inch is good enough. If more accuracy is required, you choose a ruler with finer spacing. The fact that these units are etched on the side of a straightedge implies that the procedure of measurement is "shortest, straight line between two points." You align the ruler between two points and count the tic marks. There, that's it—simple, satisfying, and comfortable.

But what is a ruler? Actually, it is just one row of an implied grid you have placed over the map. In essence, the ruler forms a reference grid, with the distance between each tic mark forming one side of a grid cell. You simply align the imaginary grid and count the cells along the row. That's easy for you, but tough for a computer. To measure distance like this, the computer would have to recalculate a transformed reference grid for each measurement. Pythagoras anticipated this potential for computer abuse several years ago and developed the Pythagorean Theorem. The procedure keeps the reference grid constant and relates the distance between two points as the hypotenuse of a right triangle formed by the grid's rows and columns. There, that's it—simple, satisfying, and comfortable for the high school mathematician lurking in all of us.

A Raster Approach

A GIS that's beyond mapping, however, asks not what mathematics can do for it, but what it can do for mathematics. How about a different way of measuring distance? Instead of measuring between two points, let's expand the concept of distance to that of *proximity*, the distance among sets of points. For a raster procedure, consider the analysis grid on the left side of figure 28.1. The distance from the cell in the lower-right corner (column 6, row 6) to each of its three neighbors is either 1.000 grid space (orthogonal) or 1.414 grid spaces (diagonal). Similar to the tic marks on your ruler, the analysis grid spacing can be small for the exacting types or fairly coarse for the rest of us.

RASTER

?	?	6.242	5.828	5.414	5.000
?	5.656	5.242	4.828	4.414	4.000
6.242	5.242	4.242	3.828	3.414	3.000
5.828	4.828	3.828	2.828	2.414	2.000
5.414	4.414	3.414	2.414	1.414	1.000
5.000	4.000	3.000	2.000	1.000	0.000

VECTOR

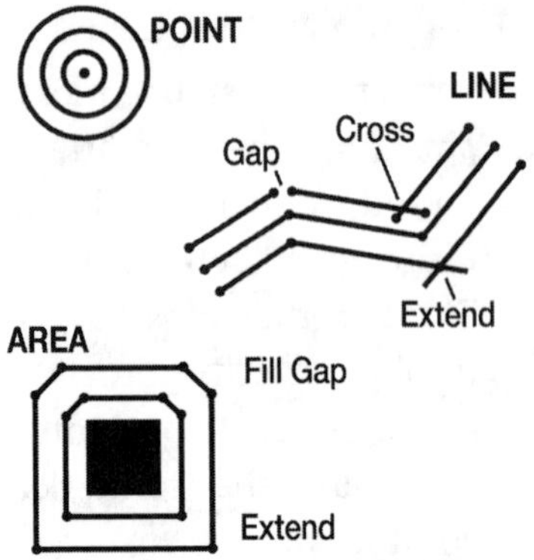

Fig. 28. 1. Characterizing simple proximity.

The distance to a cell in the next "ring" is a combination of the information in the previous ring and the type of movement to the cell. For example, position column 4, row 6 is 1.000 + 1.000 = 2.000 grid spaces, as the shortest path is orthogonal-orthogonal. You could move diagonal-diagonal passing through position column 5, row 5, as shown with the dotted line. But that route wouldn't be "shortest" because it results in a total distance of 1.414 + 1.414 = 2.828 grid spaces. The rest of the ring assignments involve a similar test of all possible movements to each cell, retaining the smallest total distance. With tireless devotion, your computer repeats this process for each successive ring. The missing information in the figure allows you to be the hero and complete the simple proximity map. Keep in mind that there are a number of possible movements from ring cells into each of the adjacent cells. (Hint: one of the answers is 7.070.)

This procedure is radically different from either your ruler or Pythagoras's theorem. It is more like nailing your ruler and spinning it while the tic marks trace concentric circles—one unit away, two units away, etc. Another analogy is tossing a rock into a still pond with the ripples indicating increasing distance. One of the major advantages of this procedure is that entire sets of "starting locations" can be considered at the same time. It's like tossing a handful of rocks into the pond; each rock begins a series of ripples. When the ripples from two rocks meet, they dissipate and indicate the halfway point. The repeated test for the smallest accumulated distance ensures that the halfway "bump" is identified. The result is a distance assignment (rippling ring value) from every location to its nearest starting location.

If your conceptual rocks represented the locations of houses, the result would be a map of the distance to the nearest house for an entire study area. Now imagine tossing an old twisted branch into the water. The ripples will form concentric rings in the shape of the branch. If your branch's shape represented a road network, the result would be a map of the distance to the nearest road. By changing your concept of distance and measurement procedure, proximity maps can be generated in an instant compared to the time you or Pythagoras would take.

However, the rippling results are not as accurate for a given unit spacing. The orthogonal and diagonal distances are exact, but the other measurements tend to overstate the true distance. For example, the rippling distance estimate for position 3,1 is 6.242 grid spaces. Pythagoras would calculate the distance as $C = ((3**2 + 5**2)) **1/2) = 5.831$. That's a difference of .411 grid spaces, or 7 percent error. As most raster systems store integer values, the rounding usually absorbs the error. But if accuracy between two points is a must, you had better forgo the advantages of proximity measurement.

Vector Considerations

A vector system, with its extremely fine reference grid, generates exact Pythagorean distances. However, proximity calculations are not its forte. The right side of figure 28.1 shows the basic considerations in generating proximity "buffers" in a vector system. First, the "reach" of the buffer is established by the user; as before, it can be small for the exacting types or fairly coarse for the rest of us. For point features, this distance serves as the increment for increasing radii of a series of concentric circles. Your high school geometry experience with a compass should provide a good conceptualization of this process. A GIS, however, evaluates the equation for a circle given the center and radius to solve for the end points of the series of line segments forming the boundary.

For line and area features, the reach is used to increment a series of parallel lines about the feature. Again, your compass work in geometry should rekindle the draftsman's approach. A GIS, on the other hand, uses the slope of each line segment and the equation for a straight line to calculate the end points of the parallel lines. "Crosses" and "gaps" occur wherever there is a bend. The intersection of the parallel lines on inside bends are clipped, and the intersection is used as the common end point. Gaps on outside bends present a different problem. Some systems simply fill the gaps with a new line segment; others extend the parallel lines until they intersect. The buffers around the square feature show that these two approaches can have radically different results. You can even take an additional step and fit a spline-fitting algorithm to smooth the lines.

A more important concern is how to account for "buffer bumping." It's only moderately taxing to calculate the series buffers around individual features, but it's a monumental task to identify and eliminate any buffer overlap. Even the most elegant procedure requires a ponderous pile of computer code and prodigious patience by the user. Also, the different approaches produce different results, affecting data exchange and integration among various systems. The only guarantee is that a stream proximity map on system A will be different from a stream proximity map on system B.

Another guarantee is that new concepts of distance are emerging as GIS goes beyond mapping. Article 29 focuses on the twisted concept of "effective" proximity, which is anything but simple and straightforward.

29 Rubber Rulers Fit Reality Better

(Calculating Effective Distance by Considering Absolute and Relative Barriers)

Rubber rulers? It must be a joke, like a left-handed wrench, a bucket of steam, or a snipe hunt. Or it could be the softening of blows in the classroom through enlightened child-abuse laws. Actually, a rubber ruler often is more useful and accurate than the old straight-edge version. It can bend, compress, and stretch throughout a mapped area as different features are encountered. After all, it's more realistic, as the straight path is rarely what most of us follow.

The previous article established the procedure for computing *simple* proximity maps as forming a series of concentric rings. The ability to characterize the continuous distribution of simple, straight-line distances to sets of features like houses and roads is useful in a multitude of applications. More importantly, the GIS procedure allows measurement of *effective* proximity, recognizing absolute and relative barriers to movement, as shown in figure 29.1. As discussed in article 28, the proximity to the starting location at 6,6 (column 6, row 6) is calculated as a series of rings. This time, however, we'll use the map on the left containing *friction factors* to incorporate the relative ease of movement through each grid cell. A friction factor of 2.00 is twice as difficult to cross as one with 1.00. Absolute barriers, such as a lake to a nonswimming hiker, are identified as infinitely far away and forces all movement around these areas.

An Exemplary Example

A generalized procedure for calculating effective distance using the friction factors is as follows (refer to fig. 29.1).

Step 1. Identify the ring cells ("starting cell" 6,6 for first iteration).

Step 2. Identify the set of immediate adjacent cells (positions 5,5; 5,6; and 6,5 for first iteration).

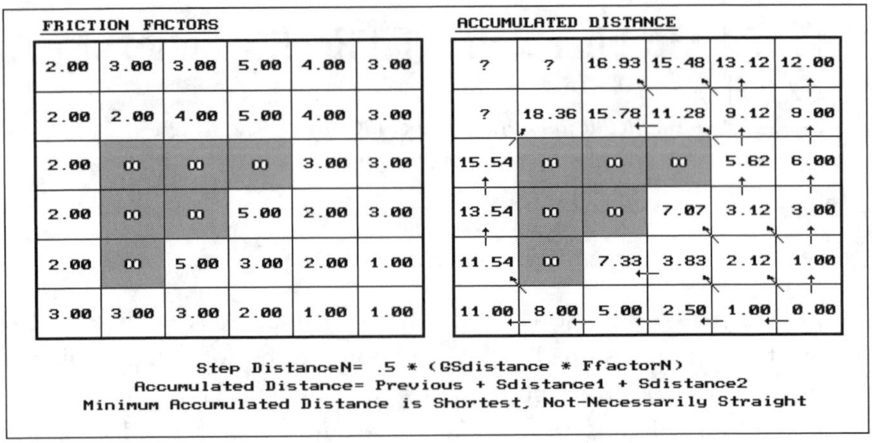

FRICTION FACTORS						ACCUMULATED DISTANCE					
2.00	3.00	3.00	5.00	4.00	3.00	?	?	16.93	15.48	13.12	12.00
2.00	2.00	4.00	5.00	4.00	3.00	?	18.36	15.78	11.28	9.12	9.00
2.00	∞	∞	∞	3.00	3.00	15.54	∞	∞	∞	5.62	6.00
2.00	∞	∞	5.00	2.00	3.00	13.54	∞	∞	7.07	3.12	3.00
2.00	∞	5.00	3.00	2.00	1.00	11.54	∞	7.33	3.83	2.12	1.00
3.00	3.00	3.00	2.00	1.00	1.00	11.00	8.00	5.00	2.50	1.00	0.00

Step DistanceN= .5 * (GSdistance * FfactorN)
Accumulated Distance= Previous + Sdistance1 + Sdistance2
Minimum Accumulated Distance is Shortest, Not-Necessarily Straight

Fig. 29. 1. Characterizing effective proximity.

Step 3. Note the friction factors for the ring cell and the set of adjacent cells (6,6 = 1.00; 5,5 = 2.00; 5,6 = 1.00; 6,5 = 1.00).

Step 4. Calculate the distance, in half-steps, to each of the adjacent cells from each ring cell by multiplying 1.000 for orthogonal or 1.414 for diagonal movement by the corresponding friction factor: .5 * (GSdirection * friction factor). For example, the first iteration ring from the center of 6,6 to the center of position

5,5 is	.5 * (1.414 * 1.00) =	.707
	.5 * (1.414 * 2.00) =	1.414
		2.121

5,6 is	.5 * (1.000 * 1.00) =	.500
	.5 * (1.000 * 1.00) =	.500
		1.000

6,5 is	.5 * (1.000 * 1.00) =	.500
	.5 * (1.000 * 1.00) =	.500
		1.000

Step 5. Choose the smallest accumulated distance value for each of the adjacent cells.

Repeat. For successive rings, the old adjacent cells become the new ring cells (the next iteration uses 5,5; 5,6; and 6,5 as the new ring cells).

That's a lot of work! Good thing you have a silicon slave to do the dirty work. Just for fun (ha!) let's try evaluating the effective distance for position 2,1. If you move from position 3,1 it's

$$.5 * (1.000 * 3.00) = \quad 1.50$$
$$.5 * (1.000 * 3.00) = \quad \underline{1.50}$$
$$3.00$$

plus previous distance = $\underline{16.93}$

equals accumulated distance =   19.93

If you move from position 3,2 it's

$$.5 * (1.414 * 4.00) = \quad 2.83$$
$$.5 * (1.414 * 3.00) = \quad \underline{2.12}$$
$$4.95$$

plus previous distance = $\underline{15.78}$

equals accumulated distance =   20.73

If you move from position 2,2 it's

$$.5 * (1.000 * 2.00) = \quad 1.00$$
$$.5 * (1.000 * 3.00) = \quad \underline{1.50}$$
$$2.50$$

plus previous distance = $\underline{18.36}$

equals accumulated distance =   20.86

Finally, choose the smallest accumulated distance value of 19.93, assign it to position 2,1, and draw a horizontal arrow from position 3,1. Provided your patience holds, repeat the process for the last three positions (answers in the next article).

The result is a map indicating the effective distance from the starting location(s) to everywhere in the study area. If the friction factors indicate time in minutes to cross each cell, then the accumulated time to move to position 2,1 by the shortest route is 19.93 minutes. If the friction factors indicate cost of haul road construction in thousands of dollars, then the total cost to construct a road to position 2,1 by the least cost route is $19,930. A similar interpretation holds for the proximity values in every other cell.

Added Realism

To make the distance measurement procedure even more realistic, the nature of the "mover" must be considered. The classic example is when two cars start moving toward each other. If the cars travel at different speeds, the geographic midpoint along the route will not be the location the cars actually meet. This disparity can be accommodated by assigning a "weighting factor" to each starter cell indicating its relative movement nature; a value of 2.00 indicates a mover that is twice as "slow" as a 1.00 value. To account for this additional information, the basic calculation in Step 4 is expanded: .5 * (GSdirection * friction factor * weighting factor). Under the same movement direction and friction conditions, a "slow" mover takes longer to traverse a given cell. Or, if the friction is in dollars, an "expensive" mover costs more to traverse a given cell (e.g., paved vs. gravel road construction).

I bet your probing intellect already has taken the next step: dynamic effective distance. We all know that real movement involves a complex interaction of direction, accumulation, and momentum. For example, a hiker walks slower up a steep slope than down it. And, as the hike gets longer and longer, all but the toughest slow down. If a long, steep slope is encountered after hiking several hours, most of us interpret it as an omen to stop for quiet contemplation.

The extension of the basic procedure to dynamic effective distance is still in the hands of GIS researchers. Most of the approaches use a "look-up table" to update the friction factor in Step 4. For example, under ideal circumstances you might hike three miles an hour in gentle terrain. When a "ring" encounters a steep adjacent cell (indicated on the slope map) and the movement is uphill (indicated on the aspect map), the normal friction is multiplied by the "friction multiplier" in the look-up table for the "steep-up" condition. This might reduce your pace to one mile per hour. A three-dimensional table can be used to simultaneously introduce fatigue—the "steep-up-long" condition might equate to zero miles per hour.

See how far you have come? From the straightforward interpretation of distance ingrained in your ruler and Pythagoras's theorem, to the twisted movement around and through intervening barriers. This bizarre discussion should confirm that GIS differs more from traditional mapping than it resembles it. Article 30 discusses how the shortest, but not necessarily straightest path, is identified.

30 Twists and Contortions Lead to Connectivity

(Identifying Optimal Paths and Routing Corridors)

❖　　❖　　❖

The last two articles challenged the assumption that all distance measurement is the "shortest straight line between two points." The concept of *proximity* relaxed the "between two points" requirement. The concept of *movement*, through absolute and relative barriers, relaxed the "straight line" requirement. What's left? Shortest, but not necessarily straight, and often among sets points.

Where does all this twisted and contorted logic lead? That's the point, *connectivity*. You know, "the state of being connected," as Webster would say. Since the rubber ruler algorithm in the previous article relaxed the simplifying assumption that all connections are straight, it seems fair to ask, "Then what is the shortest route if it isn't straight?" In terms of movement, connectivity among features involves the computation of *optimal paths*. It all starts with the calculation of an "accumulation surface," like the one shown on the left side of figure 30.1. This is a three-dimensional plot of the accumulated distance table you completed in article 29. Remember? Your homework involved that nasty, iterative, five-step algorithm for determining the friction factor weighted distances of successive rings about a starting location. Whew! The values floating above the surface are the answers to the missing table elements: 17.54, 19.54, and 19.94. How did you do?

But that's all behind us. By comparison, the optimal path algorithm is a piece of cake; just choose the steepest downhill path over the accumulated surface. All of the information about optimal routes is incorporated in the surface. Recall that as the successive rings emanate from a starting location, they move like waves bending around absolute barriers and shortening in areas of higher friction. The result is a "continuously increasing" surface that captures the shortest distance as values assigned to each cell.

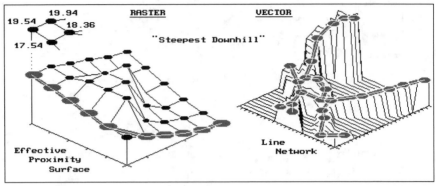

Fig. 30.1. Determining optimal paths.

In the raster example shown in figure 30.1, the steepest downhill path from the upper-left corner (position 1,1) moves along the left side of the "mountain" of friction in the center. The path is determined by successively evaluating the accumulated distance of adjoining cells and choosing the smallest value. For example, the first step could move to the right (to position 2,1) from 19.54 to 19.94 units away. But that would be stupid because it is farther away than the starting position itself. The other two potential steps, to 18.36 or 17.54, make sense, but 17.54 makes the most sense because it gets you a lot closer. So you jump to the lowest value at position 1,2. The process is repeated until you reach the lowest value of 0.0 at position 6,6.

Say, that's where we started measuring distance. Let's get this right. First, you measure distance from a location (effective distance map), then you identify another location and move downhill like a rain drop on a roof. Yep, that's it. The path you trace identifies the route of the distance wavefront (successive rings) that got there first—shortest. But why stop there when you can calculate optimal path density? Imagine commanding your silicon slave to compute the optimal paths from all locations down the surface while keeping track of the number of paths passing through each location. Like gullies on a terrain surface, areas of minimal impedance collect a lot of paths. Ready for another step? Consider weighted optimal path density. In this instance, you assign an importance value (weight) to each starting location; and, instead of merely counting the number of paths through each location, you sum the weights.

For the techy types, the optimal path algorithm for raster systems should be apparent. It's just a couple of nested loops that allow you to test for the biggest downward step of "accumulated distance" among the eight neighboring cells. You move to that position and repeat. If two or more equally optimal steps should occur, simply move to them. The algorithm stops when there aren't any more downhill steps. The result is a series of cells that forms the optimal path from the specified "starter" location to the bottom of the accumulation surface. Optimal path density requires you to build another map that counts the number of paths passing through each cell. Weighted optimal path density sums the weights of the starter locations, rather than simply counting them.

The vector solution is similar in concept, but its algorithm is a bit more tricky to implement. In the aforementioned discussion, you could substitute the words *line segment* for *cell* and not be too far off. First, you locate a starting location on a network of lines. The location might be a fire station on a particular street in your town. Then you calculate an "accumulation network" in which each line segment end point receives a value indicating shortest distance to the fire station along the street network. To conceptualize the process, the raster explanation in article 28 uses rippling waves from a tossed rock in a pond. This time, imagine waves rippling along a canal system. They are constrained to the linear network, with each point being farther away than the one preceding it. The right side of figure 30.1 shows a three-dimensional plot of this effect. It looks a lot like a roller coaster track with the bottom at the fire station (the point closest to you). Now locate a line segment with a "house on fire." The algorithm hops from the house or from "lily pad to lily pad" (line segment end points), always choosing the smallest value. As before, this "steepest downhill" path traces the wavefront that got there first—shortest route to the fire. Similarly, the concepts of optimal path density and weighted optimal path density from multiple starting locations remain intact.

What makes the vector solution testier is that the adjacency relationship among the lines is not as neatly organized as in the raster solution. This relationship, or *topology*, describing which cell abuts which cell is implicit in the matrix of numbers. On the other hand, the topology in a vector system must be stored in a database. A distinction between a vertex (point along a line) and a node (point of intersecting lines) must be maintained (see the introduction for a discussion of these terms). These points combine to form chains that, in a cascading fashion, relate to one another. Ingenuity in database design and creative use of indices and pointers for quick access to the topology are what separates one system from another. Unfettered respect should be heaped upon the programming wizards that make all this happen.

However, regardless of the programming complexity, the essence of the optimal path algorithm remains the same; measure distance from a location (effective distance map), then locate another location and move downhill. Impedance to movement can be absolute, relative barriers such as one-way streets, no left turns, and speed limits. These "friction factors" are assigned to the individual line segments and used to construct an accumulation distance network similar to that discussed in article 29. In a vector system, however, movement is constrained to an organized set of lines, instead of an organized set of cells.

Optimal path connectivity isn't the only type of connection between map locations. Consider narrowness, the shortest cord connecting opposing edges. Like optimal paths, narrowness is a two-part algorithm based on accumulated distance. For example, to compute a narrowness map of a meadow, your algorithm first selects a "starter" location within the meadow. It then calculates the accumulated distance from the starter until the successive rings have assigned a value to each of the meadow edge cells. Now choose one of the edge cells and determine the "opposing" edge cell that lies on a straight line through the starter cell. Sum the two edge cell distance values to compute the length of the cord. Iteratively evaluate all of the cords passing through the starter cell, keeping track of the smallest length. Finally, assign the minimum length to the starter cell as its narrowness value. Move to another meadow cell and repeat the process until all meadow locations have narrowness values assigned.

As you can imagine, this is a computer-abusive operation. Even with algorithm trickery and user limits, it will send the best of computers into "deep space" for awhile. Particularly when the user wants to compute the "effective narrowness" (nonstraight cords respecting absolute and relative barriers) of all the timber stands within a 1,000-x-1,000 map matrix. But GIS isn't just concerned with making things easy, be it for man or machine. It is for making things more realistic that leads to better decisions, one hopes. Optimal path and narrowness connectivity are uneasy concepts leading in that direction.

<table>
<tr><td>

```
┌─────┐
│     │
│ 31  │
│     │
└─────┘
```

</td><td>

Take a New Look at Visual Connectivity

</td></tr>
</table>

(Viewsheds and Visual Exposure Analysis)

Remember the warning the Beast gave Belle, "Don't go into the west wing"? Such is the byzantine world of GIS, full of unfamiliar bells and whistles that tempt our curiosity. Spatial derivative, effective distance, optimal paths, and narrowness are a few of the concepts discussed in recent articles. Let's continue our groping in the realm of the unfamiliar with visual connectivity. As before, the focus will be on the computational approach.

As an old Signal Corps officer, I am well versed in the manual procedure for determining FM-radio's line-of-sight connectivity. You plot the position of the general's tent (down in the valley near the good fishing hole) and spend the rest of the day drawing potential lines to the field encampments. You note when your drafted lines "dig into the dirt," as indicated by the higher contour values of ridges. In an iterative fashion, you narrow the placement of transmitters and repeaters until there is line-of-sight connectivity among all the troops. About then you are advised that the fishing played out, and the general's tent is being moved to the next valley.

In a mathematical context, the drafted lines are evaluating the relationship of the change in elevation (rise) per unit of horizontal distance (run). The "rise-to-run" ratio, or tangent, determines whether one location can be seen from another. Consider the line-of-sight plot shown in the left portion of figure 31.1. Points along a line are visible as long as their tangents exceed those of all intervening points (i.e., rise above). The seen locations are depicted by the shaded area (first four rings). The tangent values at the last two points are smaller than that of the fourth point, therefore, they are marked as not seen (unshaded). Now identify another "ray in space" and calculate where it digs into the dirt. That's it—a simple, straightforward adaptation of high school geometry's characterization of a familiar drafting technique.

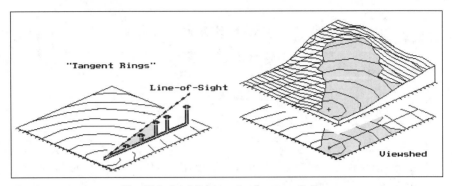

Fig. 31.1. Establishing visual connectivity.

But as we have seen, a mathematical solution can be different from a machine solution or algorithm. Just for fun, let's invoke the "rippling" wave concept used in measuring distance. "Splash," the rock falls in the lower-left corner of the map (see fig. 31.1). The elevation of the "viewer cell" is noted, along with the elevations of its eight neighboring cells. The neighboring cells are marked as seen, and eight "tangents to be beat" are calculated and stored. Attention is moved to the next ripple where the tangents are calculated and tested to determine if they exceed those on the inside ring. Locations with smaller tangents are marked as not seen. Those with larger tangents are marked as seen, and the respective tangents to be beat are updated.

This process entails just what your machine likes, neat organization. The ripple number implicitly tracks the horizontal distance. The elevations are identified by the column, row designation of the current ripple's set of cells. The tangents to be beat are identified by cell positions of the previous ripple. "Splash," followed by a series of rippling waves carrying the tangents to be beat, while the visible locations are marked in their wake. The effects of haze can be introduced by only rippling so far or adding a visual fall-off weight, such as "inverse distance squared." The lower map in the right portion of figure 31.1 is a planimetric plot of the viewshed of the same point described previously. For reference, the line-of-sight is traced on the planimetric map. The map on top drapes the *viewshed* on the elevation surface. Note that the line-of-sight shading of visible locations continues up the side of the mountain to the inflection point (ridge top) at the reference ring.

In general, the machine solution is not as exacting as a strict mathematical implementation. But what it lacks in exactness often is made up in computational speed and additional capabilities. Consider tossing a handful of rocks (viewer locations) into the pond. "Splash, splash, splash." As each ripple pattern is successively evaluated, the resultant map keeps track of the number of viewer locations connected to each map location. If 30 splashes see a location, that location is more "visually exposed" than another which has none or just a few. Instead of a viewshed map, a *visual exposure density surface* is created. It is like viewing through a fly's compound eyeball—a set of lenses (viewer cells) considered as a single viewing entity. But why stop here? Imagine the visual exposure density surface of an entire road system. High values indicate locations with a lot of windshields connected. You might term these areas "visually vulnerable" to roads. A sensitive logger might term these areas "trouble" and move the clearcut just over the ridge to areas of lower exposure values.

To make the algorithm a bit more realistic, we need to add some additional features. First, let's add the capability to discriminate among different types of viewers. For example, we all know a highway has more traffic than a back road. On the viewers map, assign the value 10 to the highway and 1 to the back road as "weighting factors." When your rippling wave identifies seen locations, it will add the weight of the viewer cell. The result is a weighted *visual exposure density surface*, with areas connected to the highway receiving higher exposure values. Isn't that more realistic? You can even simulate an aesthetics rating by assigning positive weights to beautiful things and negative weights to ugly things. A net weighted visual exposure value of zero indicates a location that isn't connected to any important feature, or has counter-balancing beauties and uglies in its visual space— far out, man!

Another enhancement is the generation of a *prominence map*. As the ripples advance, the difference between the current and the previous tangent identifies the "degree" of visual exposure. If the difference is zero or negative, the location is not seen. As the positive values increase, increasing visual prominence is indicated. A large positive difference identifies a location that is "sticking out there like a sore thumb"; you can't miss it. Couple this concept with the previous one, and you have a *weighted prominence map*. The algorithm simply multiplies the tangent difference (prominence) times the viewer weight (importance) and keeps a running sum of all positive values for each map location. The result is a map that summarizes the effective visual exposure throughout a project area to a set of map features. This is real information for decision making.

But we aren't there yet. The final group of enhancements include conditions affecting the line of sight. Suppose your viewing position is atop a 90-foot fire tower. You'd better let the computer know, so it can add 90 to the elevation value for your actual viewer height. You will see a lot more. What about the dense forest stands scattered through-out the area? You'd better let the computer know the visual screening height of each stand so it can add them to the elevation surface. You will see a lot less. Finally, suppose there is a set of features that rise above the terrain, but the elements are so thin they really don't block your vision. The elements might be a chimney, an antenna, or a pow-erline. At the instant the algorithm is testing if such a location is seen, its elevation is raised by the feature height to test if it is seen. However, the surface elevation is used in testing for tangent update to determine if any cells in the next ring are blocked. Picky, picky, picky. Sure, but while we are at it, we might as well get things right.

Figure 31.2 puts this visual analysis thing in perspective. It is a simple model determining the average visual exposure to both houses and roads for each district in a project area. The "box-and-line" flow-chart on the left summarizes the model's logic. The sentences at the bottom list the command lines implementing the processing. The visual exposure to roads and houses is determined (RADIATE), then

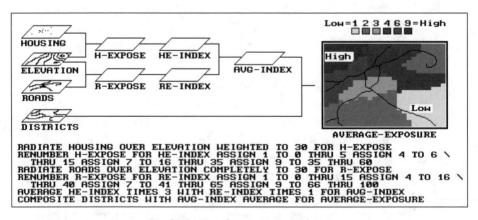

Fig. 31.2. Visual exposure model.

calibrated to a common index of relative exposure (RENUMBER). An overall index is computed (AVERAGE), then summarized for each district (COMPOSITE). The resultant map shows that the district in the upper left has a high visual exposure, not a good place for an eyesore development. But what have you learned? How would you change the model to make it more realistic? Would weighted prominence be a better measure? What about the screening effect of trees? Or the viewing distance from roads and houses? Which houses are most affected?

Recommended Reading

Books

Lowe, J., and S. Moryadas. *The Geography of Movement*, Prospect Heights, IL: Waveland Press, 1984.

Tomlin, C.D. "Characterizing Locations Within Neighborhoods." *Geographic Information Systems and Cartographic Modeling*. Englewood Cliffs, NJ: Prentice Hall, 1990.

Journal Articles

Chang, K., and B. Tsai. "The Effect of DEM Resolution On Slope and Aspect Mapping." *Cartography and Geographic Information Systems* 18: 69-77 (1991).

Lupien, A.E., et al. "Network Analysis in Geographic Information Systems." *Photogrammetric Engineering and Remote Sensing* 53(10): 1417-23 (1987).

TOPIC 10

CARTOGRAPHIC AND SPATIAL MODELING

The era of personal computing has ended. The
1990s will be the era of interpersonal computing.
—*Steve Jobs*

Many GIS applications take the technology well beyond mapping and
into the larger field of mathematical modeling. The logical sequencing
of individual map operations forms application models that express the
interrelationships among map variables. This section discusses command "macro" construction, the mathematical implications, and the
use of GIS models in consensus building and conflict resolution.

32 GIS Mirrors Perceptions of Decision Criteria

(A Structured Process from Fact to Judgment to Mapped Solution)

❖   ❖   ❖

As GIS takes us beyond mapping to application modeling, our attention increasingly focuses on the considerations embedded in the derivation of the "final" map. The map itself is valuable, but the thinking behind its creation provides the real insights for decision making. From this perspective, the model becomes even more useful than the graphic output. Yeah, sure.

No, it's true. Consider the simple model outlined in figure 32.1. It identifies the suitable areas for a campground considering basic engineering and aesthetic factors. Like any other model, it is a generalized statement, or abstraction, of the important considerations in a real-world situation. The model represents one of the most common GIS applications, a suitability model. There are other types, but for now, let's take a closer look at this one.

Exploring the Model

First, note that the model is depicted as a flowchart with boxes indicating maps and lines indicating GIS processing. It is read

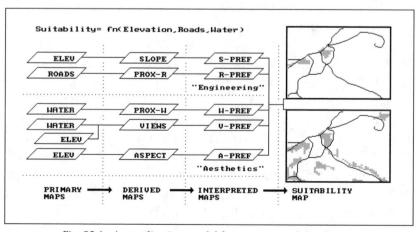

Fig. 32.1. An application model for campground development.

175

from left to right. For example, the top line tells us that a map of elevation (ELEV) is used to derive a map of relative steepness (SLOPE), which in turn is interpreted for slopes that are better for a campground (S-PREF).

Next, note that the flowchart has been subdivided into compartments by dotted horizontal and vertical lines. The horizontal lines identify separate submodels expressing suitability criteria—gently sloped, near roads, near water, with good views of water, and a westerly aspect. But more on these details later. For now concentrate on the model's overall structure. The vertical lines indicate increasing levels of abstraction. The left-most PRIMARY MAPS section identifies the base maps needed for this application. In most instances, they are physical features described through field surveys—elevation, roads, and water. They are our inventory of the landscape that we accept as "fact."

The next group is termed DERIVED MAPS. Like PRIMARY MAPS, they are "facts." However, they are difficult to collect and encode, so we use the computer to derive them. For example, slope can be measured with a clinometer, but it is impractical to collect this information for the 2,500-quarter-hectare locations (grid cells) in the project area. Similarly, the distance to roads can be measured by a survey crew "pulling tape." But it is too difficult. Note that these first two levels of model abstraction are concrete descriptions of the landscape. We could check the accuracy of our primary and derived maps simply by taking them into the field and measuring. They exist. They're tangible. We're comfortable.

The next two levels, however, are a different matter. At this juncture we move from fact to judgment—from the *description* of the landscape to the *prescription* of a proposed land use. The INTERPRETED MAPS are the result of grading landscape factors in terms of an intended use. The process involves assigning a relative "goodness" value to each map condition. For example, gentle slopes are preferred locations for campgrounds. However, if you were assessing suitability for a ski area, steeper slopes might be better. It is imperative that a common goodness scale is used for all of the interpreted maps. It's like a professor grading several exams throughout a term. Each test (*vis.* primary or derived map) is graded. And as one would expect, some students (*vis.* map locations) score well while others flunk.

The final SUITABILITY MAP is a composite of the set of interpreted maps. Like the professor at the end of the term, you simply average the test scores for each student's semester grade. There, that's it. Everyone (*vis.* each map location) is given an overall ranking. In the figure, the lower map inset identifies the best overall scores. However, you might want to do some spatial spreadsheet "what if-ing." What if views (V-PREF map) are 10 times more important than other preferences? The upper map inset shows that good locations in this scenario are severely cut back to just a few areas. But what if steepness is more important? Or proximity to water? Where are the best locations now? Are there any consistently good locations?

Model Details

Whoa! Too abstract. It's time to look at the specifics of the model. The horizontal compartments chart the processing of the individual criteria. The engineering concerns for avoiding steep slopes and large distances from existing roads are common sense. It costs a lot more to construct a campground under these conditions. Hidden behind the flowchart is the actual code (termed a *macro*) that achieves the objectives. The following macros are expressed as pMAP command sentences:

```
SLOPE ELEVATION FOR SLOPEMAP
RENUMBER SLOPEMAP FOR S-PREF ASSIGN 9 TO 0
       THRU 5 ASSIGN 8 TO 6 THRU 15 ASSIGN
       5 TO 16 THRU 25 ASSIGN 3 TO 26 THRU
       40 ASSIGN 1 TO 41 THRU 100
SPREAD ROADS TO 100 FOR PROX-R
RENUMBER PROX-R FOR R-PREF ASSIGN 9 TO 0
       ASSIGN 8 TO 1 ASSIGN 7 TO 2 THRU 3
       ASSIGN 4 TO 4 THRU 6 ASSIGN 1 TO 7
       THRU 100
```

The SLOPE and SPREAD commands create the derived maps indicating steepness and proximity to roads. These in turn are "renumbered" (i.e., calibrated) with 9 being the best and 1 being the worst. For example, 9 is assigned to the gentlest slopes of 0 percent to 5 percent, and the closest distances of 0 cells away (100-meter grid spacing).

The "aesthetic" considerations of being near water, having good views of water, and being oriented toward the west are expressed in the following sentences:

```
SPREAD WATER TO 100 FOR PROX-W
RENUMBER PROX-W FOR W-PREF ASSIGN 9 TO 0
        THRU 2 ASSIGN 7 TO 3 THRU 4 ASSIGN 4
        TO 5 THRU 6 ASSIGN 1 TO 7 THRU 100
RADIATE WATER OVER ELEVATION COMPLETELY TO
        10 FOR VIEWS
RENUMBER VIEWS FOR V-PREF ASSIGN 9 TO 30
        THRU 100 ASSIGN 8 TO 20 THRU 29
        ASSIGN 6 TO 15 THRU 19 ASSIGN 4 TO 5
        THRU 14 ASSIGN 1 TO 0 THRU 4
ORIENT ELEVATION FOR ASPECTMAP
RENUMBER ASPECTMAP FOR A-PREF ASSIGN 9 TO 6
        THRU 8 ASSIGN 7 TO 1 THRU 2 ASSIGN 3
        TO 4 THRU 5 ASSIGN 1 TO 3
```

The SPREAD, RADIATE, and ORIENT commands generate the derived maps. RENUMBER calibrates each map using the same grading scheme. For example, 9 is assigned to the closest distances to water of 0 to 2 cells away, the most visually exposed to water of 30 to 100 connections and the westerly octants of 6 to 8. Such power. You're in command. Like the professor, your interpretations control the fate of thousands of entities (*vis.* map locations).

The next step is the easy part. Just enter the following sentence for an overall ranking:

```
AVERAGE S-PREF TIMES 1 WITH R-PREF TIMES 1
        WITH W-PREF TIMES 1 WITH V-PREF TIMES
        1 WITH A-PREF TIMES 1 FOR RANKING
```

Locations with an average of 7 or better are displayed with the road network for reference in the lower inset. These locations are the contenders for the campground.

But we might want to do some additional thinking. You know, try a few things. Note that the "times 1" in the averaging command indicates the weighting factor for each map. To make good views more important, edit the sentence to "V-PREF TIMES *10*" and resubmit. The result is the map on top, with a much narrower set of choices.

Actually, there are three types of modifications you can make: weighting, calibration, and structural. Each involves editing the macro, then resubmitting. Weighting modifications affect the compositing of interpreted maps into the overall suitability map, as described previously. Calibration modifications affect the assignment of the individual "goodness" ratings. For example, you might assign 9 (best) to a broader range of slopes, say 0 percent to 10 percent. I wonder if that changes things much?

Weighting and calibration modifications are easy and straightforward—edit a parameter, resubmit, and see the effect. Structural changes are something else. They involve changing the logical structuring of the flowchart. For example, it might occur to you that forested areas are better than open terrain. To handle this, you need to add a new sequence of maps to the "aesthetics" compartment beginning with a cover-type map. Now you are GISing—conceptualizing the important considerations as maps and expressing their relationships as GIS commands. Actually that's map-ematical modeling, a piece of cake.

33 Effective Standards Required to Go Beyond Mapping

(The Implications of Geographic, Algorithm, and Interpretation Standards)

> One of the best things about standards is that there are so many different ones.
>
> —*Anonymous*

The previous article outlined the basic concepts in GIS modeling. Behind each complex map there is a sequence of commands (a macro) that reflects the "rational thinking" of the application model. The processing often is summarized into a flowchart for easier communication. Once structured, the model can be repeatedly executed in a manner similar to running "what if" scenarios in spreadsheet analysis. That must be it—GIS is merely a spatial spreadsheet.

Yep, you're right, in part. Actually, spreadsheet analysis is just one piece of a larger field called mathematical modeling. Now that maps are numbers, which we process with "map-ematics," it becomes apparent that GIS comes with all the rights, privileges, and responsibilities of other mathematics. First is a requirement to cloud common sense with a litany of terminology. At the risk of heated debate, let me suggest three broad types of models in GIS: the data model, the relational model, and the application model. Data and relational models describe how spatial information is developed and stored within the GIS. An example of a data model is the use of kriging to spatially interpolate a set of point measurements into a continuous surface, or mapped variable. Once a "map variable" is defined, the relational model assigns "spatial topology" and "attribute characteristics" within the context of the GIS.

Whew! Did you survive that opaque statement? Do you know what it means? The articles in the introduction and topic 1 discuss some of the important considerations in these models. In short, the data and relational models describe the "what and where" of spatial information. An application model, on the other hand, address-es the "so what" aspects of mapped data. Such a model investigates the intra- and interrelationships of maps. The application model is used to gain conceptual clarity and better understanding of a system or issue.

In the broadest of definitions, there are two types of application models: cartographic and spatial. The distinction between the two lies along a continuum extending from conceptual to system modeling. The degree of mathematical rigor is a good litmus test of the two types. For example, I recently had a graduate class nearly split between civil engineers and natural resource managers. The term projects of the resource managers leaned toward cartographic models expressing their understanding of an issue, such as spotted owl habitat. These concep-tual models were heavy on insight, but relatively light on mathematics and empirical study. The engineers' models, however, generally involved the spatial evaluation of existing equations, such as Horton overland flow of surface water. One student even had a model with a single equation that exceeded four lines of code, for example, $(\ln(MAP1 ** MAP2))$. The differences in approach and GIS require-ments between the cartographic and spatial modelers were apparent.

These differences, along with those raised in data and relational models, place new demands on map standards. Like the fabled Kracken in Greek mythology, standards will rise from a sea of confusion and inundate our feeble structures of paper map standards. The assault is on four fronts: exchange, geographic, algorithm, and interpretation.

Basic Standards

Exchange standards. The easiest to address, this merely involves establishing data formats for importing and exporting maps among different GISs. In the United States, the recently released spatial data transfer standard (SDTS) has made exchange standards a nonissue. But what of the other three areas of concern?

Geographic standards. Manually prepared map standards have been evolving for hundreds of years. Historically, they have been concerned with the spatial precision used in locating the boundaries of map fea-tures. Concepts, such as map scale and projection, are well developed and standardized. For the most part, these standards are translated eas-ily into the digital world of GIS. But there are some hidden pitfalls in GIS's characterization of mapped data.

A major problem lies in the assignment of numbers (thematic values) to represent the various characteristics and conditions of a map variable. For example, a soils map might contain numbers that merely reflect the color pallet used to plot the standard colors associated with each soil class. These numbers may be sufficient for most mapping and database management applications, but modeling is more demanding. Numbers from 0 to 100 might be used to identify the clay content of each soil class. For runoff modeling, a saturation index might be a more useful expression of soil distribution than simply soil class number. On a vegetation map, numbers incorporating the range of age and stocking, as well as species, might be required. A sophisticated spatial timber supply model will require a statistical description of the variance in all of these data.

That's the problem; the simple translation of map symbols and colors into numbers may not be sufficient for many of the application models. Review of geographic standards for the "corporate database" needs to be extended to include informational content, as well as locational precision. In the United States, the standards for various base maps are under review in the "national map-down," which is coordinated by the USGS. Chances are the soil and vegetation maps resulting from this review will be radically different from the current paper product expressions. GIS modeling demands will be at the core of these changes.

Algorithm Standards

The processing capabilities within a GIS also must be addressed. At the computational level, various algorithms need to be benchmarked, and users need to be given guidelines for their appropriate use. For example, the differences among maximum, average, and fitted slope algorithms should be established; users should be advised which is most appropriate for particular applications. Spatial interpolation, distance measurement, visual analysis, and fragmentation indices are other examples of algorithms awaiting review.

At another level, the processing structure of GIS can be made more standard. In the early years of database management, various products had little to do with one another. The advent of the Standard Query Language (SQL) greatly added to the utility of these systems. In a similar vein, a GIS Standard Language (GSL) would stimulate the development and exchange of application models. Without it (or at least a basic set of functionality), our modeling efforts are atomized. It's like each car company deciding where to put the clutch, brake, and gas pedals—both dumb and dangerous.

A coordinated assault on algorithm standards is not yet in place. However, several factors in the natural maturation of GIS are contributing to refinement. Within academia, the growing number of GIS courses and texts are contributing to defining a common, comprehensive processing structure. As GIS vendors look over their shoulders at the competition, they tend to incorporate the "good ideas" of others. Finally, as an increasing number of large procurements hit the street, their specifications provide a de facto definition of processing capabilities. The same maturation progression was evident in database management; GIS is just its younger sibling.

Interpretation Standards

Let's see. Exchange standards have been addressed, geographic standards are being addressed, and algorithm standards are gleams in the eyes of a venturesome few. But what about standards in the models themselves? Such concerns, referred to as *interpretation standards*, have received minimal attention. To date, emphasis has been on producing products, not on verifying the results or logic behind a final map. As more and more "modeled" maps surface, there is an increasing opportunity to scrutinize modeling results. If an area is classified as excellent elk habitat or ancient forest—but those on the ground know different—the product eventually will be deemed substandard.

Two procedures might accelerate this process. First, empirical verification results could be included with a final map, like the geographic descriptors of scale and projection. If "ground truth" shows that ancient forest was incorrectly identified one-third of the time, the user of the product should be advised. If empirical verification isn't possible, error propagation modeling can be used to estimate the reliability of the final map (see articles 19 and 20). Keep in mind that, by definition, modeling is an abstraction of reality (an "educated" guess).

A second useful tool in establishing interpretation standards is the map "pedigree." This is a new addition to a map's legend brought on by GIS modeling. In its simplest form, the pedigree is merely a listing of the macros (commands) used to create the final map. More elegant renderings also contain a flowchart of processing. These succinct descriptions of model logic provide an entry point for evaluating the model and suggesting changes. As GIS modeling matures, a map without its pedigree will be as unacceptable as a dog show contestant without its AKC papers.

In the past, maps were accepted principally on face value. A neatly drafted map indicated the cartographer's concern for accuracy. If it looked good, it probably was good. But GIS modeling has changed the playing field, as well as the rules. Without effective standards that address this new environment, GIS will have difficulty going beyond mapping.

34 Maps Speak Louder Than Words

(GIS Modeling in Consensus Building)

❖   ❖   ❖

By their nature, all land use plans contain (or imply) a map. The issue is "What should go where?" As noted in articles 32 and 33, there is a lot of thinking that goes into a final map recommendation. One simply can't arm a survey crew with a "land-use-o-meter" to measure the potential throughout a project area. The keys behind a land-use model and its interpretation by different groups are the basic elements leading to an effective land-use map solution. The map itself is merely one rendering of the thought process.

The potential of interactive GIS modeling extends far beyond its technical implementation. It promises to radically alter the decision-making environment itself. A case study might help verify this claim. The study uses three separate spatial models for allocating alternative land uses of conservation, research, and residential development. In the study, GIS modeling is used in consensus building and conflict resolution to derive the best combination of competing uses of the landscape.

Divine Mapping

The study takes place in consulting heaven—the western tip of the Caribbean island of St. Thomas. Base maps of roads, shoreline, elevation, and current flow formed the basis of the application. Separate suitability models were developed for three alternative land uses: conservation, research, and development. The final model addressed the best allocation of land by simultaneously considering all three potential landscape uses. The departure from "traditional" analysis is that the GIS was used in "real time" to respond to the questions and concerns of decision makers. In doing so, the modeling contributed to consensus building and conflict resolution, as well as to graphic portrayal of the final plan. But first, let's quickly consider the three alternative models.

A map of accessibility to existing roads and the coastline formed the basis of the conservation areas model. In determining access, the slope of the intervening terrain was considered. The slope-weighted proximity from the roads and from the coastline were calculated. In these calculations, areas that appear geographically near a road may be much less accessible. For example, the coastline may be a "stone's throw away" from the road, but it may be effectively inaccessible for recreation if it's at the foot of a cliff.

The two maps of weighted proximity from both the roads and the coast were combined into an overall accessibility map. The final step of the analysis involved interpreting relative access into conservation uses (fig. 34.1). Recreation was identified for those areas near both roads and the coast. Intermediate access areas were designated for limited use. Areas effectively far from roads were designated as preservation areas.

The characterization of the Research Areas Model first used the elevation map to identify individual watersheds. The set of all watersheds was narrowed to three based on scientists' requirements that they be relatively large and wholly contained areas (fig. 34.2). A submodel used the prevailing current to identify coastal areas influenced by each of the three terrestrial research areas.

The Development Areas Model determined the best locations for residential development. The model structure used is nearly identical to

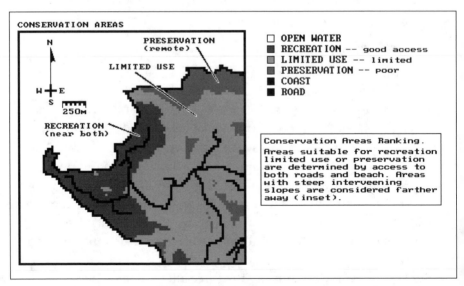

Fig. 34.1. Conservation areas map.

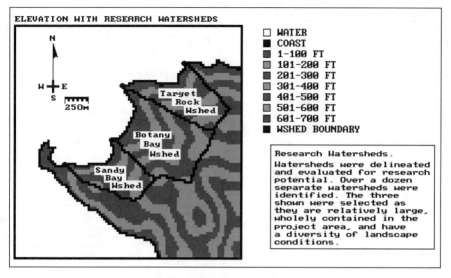

Fig. 34.2. Research areas map.

the "campground" suitability model described in article 32—a megabucks estate simply replaces a tent city. Engineering, aesthetic, and legal factors were considered. As in the classroom example in article 32, the engineering and aesthetic considerations were treated independently, as relative rankings (*vis.*, midterm test scores). An overall ranking (*vis.*, term grade) was assigned as the weighted average of the five "preference" factors. The legal constraints, however, were treated as "critical" factors. For example, an area within the 100-meter setback was considered unacceptable, regardless of its aesthetic or engineering rankings.

Putting It All Together

Figure 34.3 shows a composite map containing the simple arithmetic average of the five separate preference maps used to determine development suitability. The constrained locations mask these results and are shown as light gray (values within constrained areas are assigned the preference value of zero). Note that approximately half of the land area is ranked as "acceptable" or better (darker tones). In averaging the five preference maps, all criteria were considered equally important at this step.

The analysis was extended to generate a series of weighted suitability maps. Several sets of weights were tried. The group finally decided on

- View preference times 10 (most important),

- Coast proximity times 8,

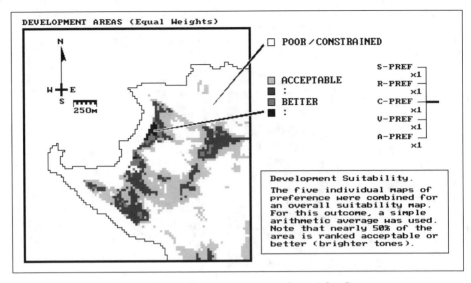

Fig. 34.3. Development areas map (unweighted).

- Road proximity times 3,

- Aspect preference times 2, and

- Slope preference times 1 (least important).

The resulting map of the weighted averaging is presented in figure 34.4. Note that a smaller portion of the land is ranked as "acceptable" or better. Also, note that the spatial distribution of these prime areas is localized to three distinct clusters.

A group of decision makers was involved in constructing all three of the individual models—conservation, research, and development. While looking over the shoulder of the GIS specialist, they saw their concerns translated into map images. They discussed whether their assumptions made sense. Debate surrounded the weights and calibrations of the models. They saw the sensitivity of each model to changes in its parameters. In short, they became involved and understood the map analysis taking place. That's a far cry from viewing a solution map for the first time at a public hearing or hearing experts reference a three-volume report each time there is a question. Heck, you didn't have the time (nor expertise) to read the report in the first place. Damned if you will read it after tonight's vote.

That's the new twist GIS modeling brings. It enables decision makers to be decision makers, not choice choosers constrained to a few

predefined alternatives. The involvement of decision makers in the
analysis process contributes to *consensus building.* As you see your concerns, and those of others, incorporated into the analysis, you get a
better feeling about the issue. In this case, the group reached consensus on the three independent land-use renderings. That sets the stage
for the final showdown: *conflict resolution.*

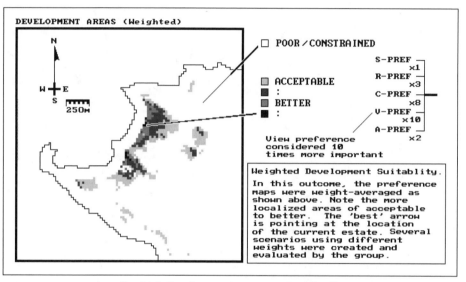

Fig. 34.4. Development areas map (weighted).

35 Is Conflict Resolution an Oxymoron?

(GIS Modeling in Conflict Resolution)

The previous article may have left you hanging. Three separate analytic approaches are used to determine the best use of a project area considering conservation, research, and development criteria. But what about areas common to two or more of the maps? These are the conflict areas in which the decision makers must either "fish or cut bait." Three basic approaches in resolving conflicts are at your disposal: hierarchical dominance, compatible use, and tradeoff. *Hierarchical dominance* assumes certain land uses are more important and, therefore, supersede all other potential uses. *Compatible use* identifies harmonious uses and assigns several uses to a single location. *Tradeoff* recognizes hardcore conflicting uses on a parcel-by-parcel basis and attempts to resolve which land use takes precedence. Effective land-use decisions involve elements of all three approaches.

Hierarchical Approach

From a map processing perspective, the hierarchical approach is easily expressed in a quantitative manner and results in a deterministic solution. Once the political system has identified a superseding use, it is relatively easy to map these areas and assign a value indicating the desire to protect them from other uses. Multiple use also is technically simple from a map analysis context, though often difficult from a policy context. When compatible uses are identified, a unique value identifying both uses is simply assigned to all areas with the joint condition.

Conflict arises when the uses are incompatible. In these instances, quantitative solutions to land-use allocation are difficult, if not impossible, to implement. The complex interaction of the frequency and juxtapositioning of several competing uses is still most effectively dealt with by human intervention. GIS technology assists decision making by deriving a map that indicates the set of alternative uses vying for each location. Once land-use information is in this graphic form, decision makers can assess the patterns of conflicting uses and determine land-use allocations. GIS also can assist by comparing different allocation scenarios and identifying areas of difference.

In the St. Thomas study in article 34, the hierarchical dominance approach was tried, but it was a total failure. At the onset, the group was uncomfortable with identifying one land use as always better than another. Just for fun, however, the approach was demonstrated by identifying development as least favored, recreation next, and the researchers' favorite watershed taking final precedence. The resulting map was rejected as it contained little area for development, and the areas that were available were scattered into disjointed parcels—unfeasible conditions. Even if you could clarify conflict in "policy space," it is frequently muddled in the complex reality of geographic space.

Conflicts Map

The alternative approaches of compatible use and tradeoff depend on generating a map indicating all of the competing land uses for each location in a project area, a comprehensive *conflicts map*. Figure 35.1 is such a map, considering the conservation areas, research areas, and development areas maps. Note that most of the area is without conflict (lightest tone). In the absence of spatial guidance in a conflicts map, there is a tendency to assume every square inch is in conflict. In the presence of a conflicts map, however, attention is quickly focused on the unique patterns of actual conflict.

First, the areas of actual conflict were reviewed for compatibility. For example, it was suggested that research areas could support limited-use hiking trails, and both activities were assigned to those locations. However, most of the conflicts were real and had to be resolved "the hard way." Figure 35.2 presents the group's "best" allocation of land use. Dialogue and group dynamics dominated the tradeoff process. As in all discussions, individual personalities, persuasiveness, rational arguments, and facts affected the collective opinion. The initial breakthrough was the agreement that the top and bottom research areas should remain intact. In part, this made sense as these areas had significantly less conflict than the central watershed.

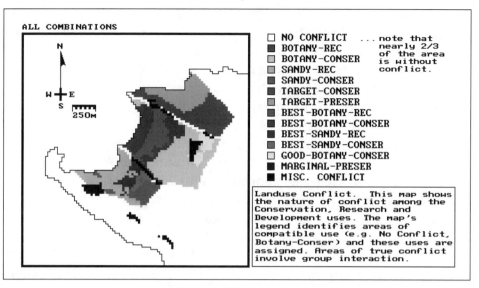

Fig. 35.1. Conflicts map.

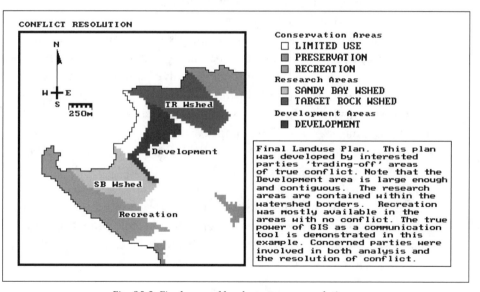

Fig. 35.2. Final map of land-use recommendations.

Spatial Dialogue

It was decided that all development should be contained within the central watershed. Structures would be constrained to the approximately 20 contiguous hectares identified as best for development, which was consistent with the island's policy to encourage "cluster" development. The legally "constrained" area between the development cluster and the coast would be for the exclusive use of the residents. The adjoining research areas would provide additional buffering and open space, thereby enhancing the value of the development. In fact, it was pointed out that this arrangement provided a third research setting to investigate development, with the two research watersheds serving as control.

Recreation use then received the group's attention. This step was easy, because a large part of the best recreation area was in the southern portion and reflected minimal conflict with the other uses. Finally, the remaining small "salt-and-pepper" parcels were absorbed by their surrounding "limited or preservation use" areas. In all, the group's final map is a fairly rational land-use allocation result and one that is readily explained and justified. Although the decision group represented several diverse opinions, the final map achieved consensus. In addition, each person felt as though he or she actively participated and, by using the interactive process, better understood both the area's spatial complexity and the perspectives of others.

This last step of tradeoffs in the analysis may seem anticlimactic. After a great deal of "smoke and dust raising" about computer processing, the final assignment of land uses involved a large amount of subjective judgment. This point, however, highlights the capabilities and limitations of GIS technology. GIS provides significant advances in how we manage and analyze mapped data. It rapidly and tirelessly allows us to assemble detailed spatial information. GIS also allows us to incorporate more sophisticated and realistic interpretations of the landscape. It doesn't, however, provide an artificial intelligence for land-use decision making. GIS technology greatly enhances our decision-making capabilities, but it does not replace them. It is both a toolbox of advanced analysis capabilities and a sandbox to express our creativity and concerns.

Recommended Reading

Books

Berry, J.K. "GIS in Island Resource Planning: A Case Study in Map Analysis." In *Geographical Information Systems: Principles and Applications*, ed. D.J. Maguire, M.F. Goodchild, and D.W. Rhind, Vol. 2, 285-95. Essex, UK: Longman, 1991.

———. "GIS Resolves Land-Use Conflicts: A Case Study." *1993 International GIS Sourcebook*, 248-53. Fort Collins, CO: GIS World, 1992.

Calkins, H.W. "GIS and Public Policy." In *Geographical Information Systems: Principles and Applications*, ed. D.J. Maguire, M.F. Goodchild, and D.W. Rhind, Vol. 2, 233-45. Essex, UK: Longman, 1991.

Dangermond, J. "The Future of GIS Technology." *The 1990 GIS Sourcebook*, 7-11. Fort Collins, CO: GIS World, 1990.

Densham, P.J. "Spatial Decision Support Systems." In *Geographical Information Systems: Principles and Applications*, ed. D.J. Maguire, M.F. Goodchild, and D.W. Rhind, Vol. 1, 403-12. Essex, UK: Longman, 1991.

Guptill, S.C. "Spatial Data Exchange and Standardization." In *Geographical Information Systems: Principles and Applications*, ed. D.J. Maguire, M.F. Goodchild, and D.W. Rhind, Vol. 1, 515-30. Essex, UK: Longman, 1991.

Robinson, V., et al. "Expert Systems for Geographic Information Systems in Natural Resource Management." In *Fundamentals of Geographic Information Systems: A Compendium*, ed. W. Ripple, 155-65. Bethesda, MD: American Society of Photogrammetry and Remote Sensing, 1989.

Star, J., and J. Estes. "Looking Toward the Future." Chapt. 13 in *Geographic Information Systems: An Introduction*. Englewood Cliffs, NJ: Prentice Hall, 1990.

Strand, E.J. "A Profile of GIS Standards." *1991-92 International GIS Sourcebook*, 417-21. Fort Collins, CO: GIS World, 1991.

Tom, H. "Geographic Information Systems Standards: A Federal Perspective." *The 1990 GIS Sourcebook*, 281-84. Fort Collins, CO: GIS World, 1991.

———. "Spatial Information and Technology Standards Evolving." *1991-92 International GIS Sourcebook*, 422-24. Fort Collins, CO: GIS World, 1991.

Tomlin, C.D. "Descriptive Modeling" and "Prescriptive Modeling." *Geographic Information Systems and Cartographic Modeling*. Englewood Cliffs, NJ: Prentice Hall, 1990.

Journal Articles

Andrew, F., et al. "A Perspective on GIS Technology in the Nineties." *Photogrammetric Engineering and Remote Sensing* 57(11): 1431-37 (1991).

Archer, H., and P. Croswell. "Public Access to Geographic Information Systems: An Emerging Legal Issue." *Photogrammetric Engineering and Remote Sensing* 55(11): 1575-82 (1989).

Berry, J.K., and J. Berry. "Assessing Spatial Impacts of Land-Use Plans." *Journal of Environmental Management* 27: 1-9 (1988).

Chrisman, N.R. "Design of Geographic Information Systems Based on Social and

Cultural Goals." *Photogrammetric Engineering and Remote Sensing* 53(10): 1367-70 (1987).

Demko, G.J. "Geography Beyond the Ivory Tower." *Annals of the Association of American Geographers* 78(4): 575-79 (1988).

Gimblett, H.R. "Visualizations: Linking Dynamic Spatial Models with Computer Graphics Algorithms for Simulating Planning and Management Decisions." *Journal of Urban and Regional Information Systems Association* 2(2): 26-34 (1990).

Johnston, C., et al. "Geographic Information Systems for Cumulative Impact Assessment." *Photogrammetric Engineering and Remote Sensing* 54(11): 1609-15 (1988).

Mark, M., and M. Gould. "Interacting with Geographic Information: A Commentary." *Photogrammetric Engineering and Remote Sensing* 57(11): 1427-36 (1991).

Epilog

❖ ❖ ❖

I would have written a shorter book, if I had time.
–Mark Twain (paraphrased)

Geographic information system (GIS) technology has grown up and is about to face the utilitarian user who lacks the sentimental attachment of the earlier GIS zealots. The excitement of "developing a technology for technology's sake" is rapidly being domesticated and put to practical use. However, to more fully understand the modern GIS, we have to follow the King of Heart's advice to Alice, "Begin at the beginning, and go on until you come to the end; then stop."

Historical Overview

Information has always been the cornerstone of effective decisions. Resource and environmental information are particularly complex because they require two descriptors: WHERE IS WHAT. For hundreds of years the link between the two descriptors has been the traditional, manually drafted map. Its historical use was for navigation through unfamiliar terrains and seas with emphasis on the accurate placement of physical features.

More recently, analysis of mapped data for decision making has become an important part of resource planning. This new perspective marks a turning point in the use of maps—from`emphasizing physical descriptions of geographic space, to interpreting mapped data, and, finally, to spatially communicating decision factors. This movement from WHERE IS WHAT to SO WHAT has set the stage for entirely new concepts in planning and management.

Since the 1960s, the decision-making process has become increasingly quantitative, and mathematical models have become commonplace. Prior to the computerized map, most spatial analysis concepts were limited in their practical implementation. The computer has provided the means for both efficient handling of voluminous data and effective spatial analysis capabilities. From this perspective, all GISs are rooted in the digital nature of the computerized map. Consider the rhetorical (and tongue-in-cheek) question, "What problems do paper maps uniquely solve? Decoration and fire starting."

Computer Mapping

The early 1970s saw *computer mapping* automate the cartographic process. The points, lines, and areas defining geographic features on a map are represented as an organized set of x,y coordinates. These data form input to a pen plotter that can rapidly redraw the connections at a variety of scales and projections. The map image is the focus of this process. In one sense, you can say GIS is based on "LAT/LONG"—look at that (LAT) and lots of neat graphics (LONG).

But that's an unfair statement. The pioneering work during this period established many of the underlying concepts and procedures of modern GIS technology. An obvious advantage of computer mapping is the ability to change a portion of a map and quickly redraft the entire area. Revisions to resource maps, such as a forest fire update, that would normally take weeks now can be done in a few hours. The less obvious advantage is the radical change in the format of mapped data—from analog inked lines on paper to digital values stored on disk. However, the most lasting implication of computer mapping is the realization that it comes with "some assembly required." Before you can map, you must assemble a deluge of digital data on your disk.

Spatial Data Management (Descriptive)

The early 1980s exploited the change in format and computer environment of mapped data. *Spatial database management systems* were developed that linked computer mapping techniques with traditional database capabilities. For example, a user is able to point at any location on a map and instantly retrieve information about that location. Or, a user can specify a set of conditions, such as a specific forest and soil combination, and direct the result of the geographic search to be displayed as a map, like a database with a picture waiting to happen.

The demand for this spatially and thematically linked data focused attention on data issues. The result has been an integrated processing environment for a wide variety of mapped data. In resource applications, it allows remotely sensed imagery, digital elevation, roads, and vegetation maps to coexist in the same computing environment. In the computer context, it sets the stage for spatial data to be fully integrated with standard office automation and electronic communication systems.

Increasing demands for mapped data focused attention on data availability, accuracy, and standards. Hardware vendors continued to improve digitizing equipment, with manual digitizing tablets giving way to automated scanners at many GIS sites. A new industry for map encoding and database design emerged, as well as a marketplace for the sales of digital map products. Regional, national, and international organizations began addressing the necessary standards for digital maps to ensure compatibility among systems. This era saw GIS database development move from project costing to equity investment justification—the ultimate form of recognition from the all-powerful auditor types.

Spatial Analysis and Modeling (Prescriptive)

Concurrent with database development, attention focused on analytical operations. The early GIS systems concentrated on automating our current mapping practices. If a resource manager had to overlay several maps on a light-table, an analogous procedure was developed within the GIS. Similarly, if repeated distance and bearing calculations were needed, the GIS was programmed with a mathematical solution. The result of this effort was GIS functionality that mimicked the manual procedures in our daily activities. The value of these systems was the savings gained by automating tedious and repetitive operations.

By the mid-1980s a comprehensive theory of spatial analysis began to emerge. To many, this theory is as uncomfortable as it is unfamiliar. However, the next time you access your powerful yet potentially dangerous GIS, keep in mind that to err is human, but it takes a computer to really mess things up.

Spatial Statistics

The "map-ematical" approach takes two forms: spatial statistics and spatial modeling. On the one hand, spatial statistics has been used by geophysicists for many years in characterizing the geographic distribution or pattern of mapped data. The statistics describe the spatial variation in the data rather than assuming a typical response is everywhere. For example, field measurements of snow depth can be made at several plots within a watershed. Traditionally, these data are analyzed for a single value (the average depth) to characterize the watershed.

On the other hand, spatial statistics uses both the location and the measurements at the plots to generate a map of relative snow depth throughout the entire watershed. The full impact of this treatment of maps is yet to be determined. The application of such concepts as spatial correlation, statistical filters, map uncertainty, and error propagation await their translation from other fields.

Spatial Modeling

The second "map-ematical" approach has a rapidly growing number of current resource applications. Journals and conference proceedings are filled with descriptions of new and exciting application models. These procedures express a resource concern as a series of map analysis steps, leading to a "solution map." For example, forest managers can characterize timber supply by considering the relative skidding and log-hauling accessibility of harvesting parcels. Wildlife managers can consider such factors as proximity to roads and relative housing density to map human activity and incorporate this information into habitats. Landscape planners can assess the visual exposure of alternative sites for a facility to sensitive viewing locations, such as roads and scenic overlooks.

Most of the traditional mathematical capabilities, plus an extensive set of advanced map processing operations, are available in modern GIS packages. You can add, subtract, multiply, divide, exponentiate, root, log, cosine, differentiate, and even integrate maps. After all, maps in a GIS are just an organized set of numbers. However, with "map-ematics," the spatial coincidence and juxtaposing of values among and within maps create new operations, such as effective distance, optimal path routing, visual exposure density and landscape diversity, and shape and pattern. To some, the fractal dimension and second derivative of a map actually have meaning. To others, General Halftrack (in the Beetle Bailey comic strip) summed it up saying that the only problem with having sophisticated equipment is that there is no one sophisticated enough to use it.

That brings me to the "Go Bang" condition. At one point in my grandfather's career, he was a district ranger in the Forest Service, and Go Bang was his horse. He would ride throughout his district and, to hear him tell, he knew every tree branch and blade of grass. Then the pickup truck arrived; from then on the forester was removed from the forest with only a windshield vignette of the place. He knew when forestry died. My father was a consulting forester who, while he cursed his old Studebaker pickup, he never entertained a romantic thought of steering a cantankerous beast through the woods. But, then again, he felt that those damn aerial photos kept the forester's head in the clouds. He knew when forestry died. So, what about this GIS thing? Is it just another layer of technology that further removes managers from the space they manage? Where is it taking us? Will future generations simply envelop GIS as it moves on to the next technological confrontation? Will we ever really know when forestry dies? Or any other spatially dependent field?

Spatial Communication (Perceptive)

Like the phoenix, technology seems to rise from ashes of the uninitiated. Increasingly, it seems we are becoming a litigious society. Each resource decision is challenged; and, as a result, the "decision impotency" syndrome is running rampant: "Ready, aim, aim, aim, aim. . . ." Effective decisions are rare indeed and the decision-making process has been infected with "paralysis through analysis" and a "data gridlock." In response to this malaise, the 1990s is building on the cognitive basis as well as the databases of GIS technology.

Resource information systems are at a threshold that is pushing us well beyond mapping, management, and even modeling to *spatial reasoning and dialogue*. In the past, analysis models have focused on management options that are technically optimal—the scientific solution. Yet in reality, there is another set of perspectives that must be considered—the social solution. It is this final sieve of management alternatives that most often confounds resource decision making. It uses elusive measures such as human values, attitudes, beliefs, judgment, trust, and understanding. These are not the usual quantitative measures amenable to computer algorithms and traditional decision-making models.

The step from technically feasible to socially acceptable options is not so much one of increased scientific and econometric modeling as it is one of communication. Basic to effective communication is involvement of interested parties throughout the decision-making process. This new participatory environment has two main elements: consensus building and conflict resolution. *Consensus building* involves technically driven communication and occurs during the alternative formulation phase. It involves the resource specialist's translation of the various considerations raised by a decision team into a spatial model. Once completed, the model is executed under a wide variety of conditions, and the differences in outcome are noted.

From this perspective, a single map of a forest plan is not the objective. It is how maps change as the different scenarios are tried that becomes information. "What if avoidance of visual exposure is more important than avoidance of steep slopes in siting a new haul road?" "Where does the proposed route change, if at all?" Answers to these analytic queries focus attention on the effects of differing perspectives. Often, seemingly divergent philosophical views result in only slightly different map views. This realization, coupled with active involvement in the decision-making process, often lead to group consensus.

However, if consensus is not obtained, *conflict resolution* is necessary. This approach ("Nobody is right if everybody is wrong") seeks an acceptable management action through the melding of different perspectives. Socially driven communication occurs during the decision formulation phase. It involves the creation of a "conflicts map" that compares the outcomes from two or more competing uses. Each management parcel is assigned a numeric code describing the actual conflict over the location. A parcel might be identified as ideal for wildlife preservation, a campground, or a timber harvest. As these alternatives are mutually exclusive; a single use must be assigned. The assignment, however, involves a holistic perspective that simultaneously considers the assignments of all other locations in a project area.

Traditional scientific approaches are rarely effective in addressing the holistic problem of conflict resolution. Even if a scientific solution is reached, it is viewed with suspicion by the layperson. Modern resource information systems provide an alternative approach involving human rationalization and tradeoffs. This process involves such statements as, "If you let me harvest this parcel, I will let you set aside that one as a wildlife preservation." The statement is followed by a persuasive argument and group discussion. The dialogue is far from a mathematical optimization but often closer to an effective decision. It uses the information system to focus discussion away from broad philosophical positions to a specific project area and its unique distribution of conditions and potential uses.

Conclusion

Geographic information system technology is evolutionary, not revolutionary. It responds to contemporary needs as much as it responds to technical breakthroughs. Planning and management have always required information as their cornerstone. Early information systems relied on physical storage of data and manual processing. With the advent of the computer, most of these data and procedures were automated. As a result, the focus of these systems has been expanded from descriptive inventories to prescriptive analysis. In this transition, map analysis has become more quantitative. This wealth of new processing capabilities provides an opportunity to address complex issues in entirely new ways.

It is clear that GIS technology has greatly changed our perspective of a map. It has moved map processing from its historical role as a provider of input to an active and vital ingredient in the "throughput" process of decision making. Today's professional is challenged to understand this new environment and formulate innovative applications that take us beyond mapping and into the twenty-first century.

Appendix A
The Tutorial Map Analysis
Package (tMAP™)

❖   ❖   ❖

Description

The Tutorial Map Analysis Package (tMAP) is an easy way to get your hands on GIS technology and the concepts, algorithms, and issues presented in this book. tMAP is designed for self-learning map analysis concepts. The software contains 10 specially designed tutorials corresponding to the 10 topics in this book. In addition, there are several other tutorials demonstrating map analysis applications. Users are shown how to encode their own data and relate their own application models.

The tMAP system is a special version of the Professional Map Analysis Package (pMAP™) commercial software and the Academic Map Analysis Package (aMAP™) educational materials for classroom study*. The tutorial software contains fully functional software, tutorial database, exercises, and brief text. tMAP provides advanced map processing analysis capabilities including:

- optimal paths • visual exposure • edge characterization

- spatial interpolation • slope/aspect • proximity

- coincidence statistics • roving window summaries

- map overlay • geographic searches • thematic mapping

- contouring • shape/pattern analyses • contiguity

- macro files • pop-up windows • contextual help

all within a user-friendly natural language. The tMAP system uses a raster data structure yet allows for input in the form of gridded data, digitized points, lines, or polygons. Output includes summary tables, character-based or rater pattern maps, monochrome or color display, and standard files for enhanced color and line graphs.

tMAP Specifications

Disk Format: 3.5", DS/HD, DOS formatted (1.44MB)
Environment: Any personal computer (PC) with a hard disk and DOS
    ver. 3.2 or higher operating system; Intel 286, 386, or 486 central
    processors; coprocessors 287 and 387 recommended but not required
Map Size: 25 rows by 25 columns (tutorial database)
Copyright © 1993 by Spatial Information Systems, Inc. All rights reserved.
    Licensed for personal, noncommercial use only.

---

*For more information about pMAP and aMAP contact Spatial Information Systems Inc., 19 Old Town Square, Fort Collins, CO 80524, phone 303-490-2155.

tMap Tutorials

The tutorials are keyed to the text and are graded from beginning through advanced levels of difficulty.

Level of Difficulty

Beginning to Intermediate
TUTOR1-9.CMD

Tutorials corresponding to the brief text (TEXT.DOC)on the tMAP diskette.

Intermediate to Advanced
TMAP0-10.CMD

Tutorials demonstrating analytic capabilities as described in *Beyond Mapping.*

Advanced

TU-ERODE.CMD	Simple erosion potential model
TU-ACT.CMD	Human activity derivation model
TU-VIEW.CMD	Visual exposure analysis model
TU-INTRP.CMD	Spatial interpolation model
TU-SED.CMD	Effective distance sediment loading model
TU-ACCES	Timber access and facilities siting model
TU-RESP.CMD	Wildfire response model
TU-RISK.CMD	Wildfire risk model
TU-WATER.CMD	Considerations in encoding points, lines, and polygons
TU-SOILS.CMD	Considerations in encoding adjacent polygons
TU-ELEV.CMD	Considerations in encoding elevation

Topic/tMAP Correlation

Introduction

TMAP0.CMD	Basic introduction to tMAP system
TUTOR9.CMD	Data entry into tMAP
TU-WATER.CMD	Considerations in encoding points, lines, and polygons

Topic 1

TMAP1.CMD	Spatial interpolation
TUTOR2.CMD	Reclassifying maps
TU-INTRP.CMD	Spatial interpolation model

Topic 2

TMAP2.CMD	Effective distance
TUTOR5.CMD	Distance measurement
TUTOR6.CMD	Connectivity operations
TU-RESP.CMD	Wildfire response model

Topic 3

TMAP3.CMD	Roving windows
TUTOR7.CMD	Neighborhood summaries

Topic 4
TMAP4.CMD   Potpourri of map analysis operations
TU-ACT.CMD   Human activity derivation model
TU-VIEW.CMD   Visual exposure analysis model

Topic 5
TMAP5.CMD   Assessing shape and pattern

Topic 6
TMAP6.CMD   Uncertainty and error propagation

Topic 7
TMAP7.CMD   Overlaying maps
TUTOR3.CMD   Point-by-point overlay
TUTOR4.CMD   Regionwide overlay
TU-ERODE.CMD   Simple erosion potential model

Topic 8
TMAP8.CMD   Analyzing power line siting
TU-RISK.CMD   Wildfire risk model

Topic 9
TMAP9.CMD   More on slope, distance, and viewsheds
TU-SED.CMD   Effective distance sediment loading model
TU-ACCES.CMD   Timber access and facilities siting model

Topic 10
TMAP10.CMD   Campground development model

Order Form

ITEM NO.	ISBN	TITLE	PRICE	AMOUNT
29.0	0-9625063-6-2	Beyond Mapping	$32.95	
29.1	1-882610-00-8	tMAP™ Software	$21.95	
29.2	1-882610-01-6	Book/Software Package	$46.75	

GIS WORLD BOOKS

Subtotal	
Colorado residents add 3% sales tax Canadian residents add 7% GST	
Shipping and handling: U.S. and Canadian orders add US$3.50/item Overseas orders US$12/item	
TOTAL	

97BB-04

Name _____

Organization _____

Address_____

City_____State/Province _____

ZIP/Mail Code_____Country_____

Phone _____

FAX _____

Sorry. Only orders prepaid in U.S. funds can be accepted. Allow 4-6 weeks for delivery.

☐ Check/Money Order Enclosed

Purchase Order Number: _____

☐ VISA   ☐ MasterCard

Credit Card Number:

☐☐☐☐☐☐☐☐☐☐☐☐☐☐☐☐

Signature _____Expiration Date _____

To order: 1. Phone: 970-221-0037   **2. FAX:** 970-221-5150
3. Mail to: GIS World Inc., 400 N. College Ave., Suite 100, Fort Collins, CO 80524, USA

Appendix B
Resources[*]

❖   ❖   ❖

Professional Societies

American Congress on Surveying and Mapping (ACSM)
5410 Grosvenor Ln.
Bethesda, MD 20814
USA
Phone: 301-493-0200
FAX: 301-493-8245
> Comprised of three member organizations: the American Cartographic Association (ACA), the American Association for Geodetic Surveying (AAGS), and the National Society for Professional Surveyors (NSPS). Jointly sponsors annual conference with American Society for Photogrammetry and Remote Sensing (ASPRS). Publishes a bimonthly news magazine (*ACSM Bulletin*) and two scholarly journals (*Surveying and Land Information Systems* and *Cartography and Geographic Information Systems*).

American Society for Photogrammetry and Remote Sensing (ASPRS)
5410 Grosvenor Ln.
Bethesda, MD 20814
USA
Phone: 301-493-0290
FAX: 301-493-0208
> A high-technology society serving such scientific communities as mapping sciences, photogrammetry, remote sensing, and GIS. Jointly sponsors annual conference with American Congress on Surveying and Mapping (ACSM) and the annual GIS/LIS conference with Association of American Geographers (AAG), Automated Mapping/Facilities Management (AM/FM) International, American Congress on Surveying and Mapping (ACSM), and the Urban and Regional Information Systems Association (URISA). Publishes a monthly journal, *Photogrammetric Engineering & Remote Sensing*.

Association for Geographic Information (AGI)
12 Great George St.
London SWIP 3AD
UNITED KINGDOM
Phone: 44 71 222 7000 ext. 226
FAX: 44 71 222 9430
> The major nongovernmental, United Kingdom umbrella organization promoting GIS technology growth and interests. Sponsors an annual meeting and several regional meetings and seminars. Publishes a year-book (*Geographic Information*) and newsletter.

---

*For more information see the *1993 International GIS Sourcebook* (Fort Collins, CO: GIS World, 1992, 413-18) or contact each society directly.

213

Association of American Geographers (AAG)
1710 Sixteenth St., N.W.
Washington, DC 20009-3198
USA
Phone: 202-234-1450
FAX: 202-234-2744

Promotes and encourages geographic research and education and disseminates research findings to education, government, and business members in North America and abroad. Sponsors an annual conference and jointly sponsors an annual GIS/LIS conference with American Society for Photogrammetry and Remote Sensing (ASPRS), Automated Mapping/Facilities Management (AM/FM) International, American Congress on Surveying and Mapping (ACSM), and the Urban and Regional Information Systems Association (URISA).

Australasian Urban and Regional Information Systems Association (AURISA)
PO Box E307
Queen Victoria Terrace, ACT 2600
AUSTRALIA
Phone: 61 6 273 4054
FAX: 61 6 273 4057

Membership ranges from local, state, and federal government authorities to public utilities , academia, vendors, and consultants in Australia, New Zealand, and the Asia-Pacific region. Sponsors an annual conference and publishes a newsletter (*AURISA News*), technical monographs, and annual conference proceedings.

Automated Mapping/Facilities Management (AM/FM) International
14456 E. Evans Ave.
Aurora, CO 80014
USA
Phone: 303-337-0513
FAX: 303-337-1001

Serves utilities; local, state, and federal government agencies; interested organizations; and the general public. Sponsors an annual conference and a bimonthly newsletter.

Automated Mapping/Facilities Management International, European Division (AM/FM ED)
PO Box 6
CH-4005 Basel
SWITZERLAND
Phone: 41 61 691 5111
FAX: 41 61 691 8189

A nonprofit educational association to advance and promote the benefits of geographic and facilities management information systems. Sponsors annual European and regional conferences and publishes a quarterly newsletter and membership directory .

Canadian Association of Geographers (CAG)
Burnside Hall
McGill University
Rue Sherbrooke St.,W.
Montreal, PQ
CANADA H3A 2K6
Phone: 514-398-4946
FAX: 514-398-7437
 Promotes geographical research and teaching and represents geographers and the profession in scientific and business communities. Sponsors an annual conference hosted by different geography departments across Canada in addition to regional meetings. Publishes two journals, *The Canadian Geographer* and *The Operational Geographer*.

Canadian Institute of Surveying and Mapping (CISM)
206-1750 rue Courtwood Crescent
Ottawa, ON
CANADA K2C 2B5
Phone: 613-224-9851
FAX: 613-224-9577
 A national surveying and mapping association. Sponsors an annual conference and publishes annual conference proceedings, the *CISM Journal*, and technical publications.

Urban and Regional Information Systems Association (URISA)
900 Second St., N.E., Ste. 304
Washington, DC 20002
USA
Phone: 202-289-1685
FAX: 202-842-1850
 Concerned with the effective use of information systems by local, state/provincial, and federal governments. Sponsors an annual conference and publishes the *Annual Conference Proceedings*, the semiannual *URISA Journal*, the *URISA News*, the *URISA Marketplace* (a monthly listing of available employment positions), and the *URISA Membership Directory*.

GIS Information Clearinghouses

National Center for Geographic Information and Analysis (NCGIA)
University of California at Santa Barbara
3510 Phelps Hall
Santa Barbara, CA 93106-4060
USA
Phone: 805-893-8224
FAX: 805-893-8617
 Consortium of three universities—University of California at Santa Barbara, University of Maine at Orono, and the State University of New York at Buffalo—that endorses standardizing guidelines for GIS curriculum and maintains a library and bibliography of articles and publications on GIS and GIS-related issues.

U.S. Geological Survey (USGS)
Earth Science Information Center (ESIC)
507 National Center
Reston, VA 22092
USA
Phone: 713-646-6045
A good source of digital line graph (DLG) data collected primarily by federal agencies. Call 800-USA-MAPS to be put in touch with one of the dozen regional ESICs around the United States.

Glossary of GIS Terms

❖ ❖ ❖

Bruce L. Kessler

This glossary presents many of the words and phrases that a geographic information system (GIS) user may encounter. Some terms have been simplified to limit the overwhelming feeling that many people get when first presented with a GIS. Also, an effort was made to eliminate bias toward any GIS software package. If there is some residual slant, the author does not intend to imply that one GIS package is better or worse than the other. Words set in boldface are defined elsewhere in the glossary. Numbers refer to the reference list at the end of the glossary.

absolute map accuracy
The accuracy of a map in relationship to the earth's geoid. The accuracy of locations on a map that are defined relative to the earth's geoid are considered absolute because their positions are global in nature and accurately fix a location that can be referenced to all other locations on the earth. Contrast *absolute map accuracy* with *relative map accuracy*.[3]

acceptance test
A set of particular activities performed to evaluate a hardware of software system's performance and conformity to specifications.

accuracy
1. If applied to paper maps or map databases, degree of conformity with a standard or accepted value. *Accuracy* relates to the quality of a result and is distinguished from **precision**.[5]
2. If applied to data collection devices such as digitizers, degree of obtaining the correct value.

address matching
A mechanism for relating two files using address as the key item. Geographic coordinates and attributes subsequently can be transferred from one address to the other.[2]

algorithm
A step-by-step procedure for solving a mathematical problem. For instance, the conversion of data in one map projection to another map projection requires that the data be processed through an algorithm of precisely defined rules or mathematical equations.

aliasing
The occurrence of jagged lines on a raster-scan display image when the detail exceeds the resolution on the screen.[1]

Bruce L. Kessler is a GIS instructor, project manager, and applications specialist for Environmental Systems Research Institute, Inc. in Boulder, Colo. The "Glossary of GIS Terms" originally appeared in the *Journal of Forestry* (November 1992), published by the Society of American Foresters. It is reprinted by permission and cannot be reproduced.

American National Standards Institute (ANSI)
An institute that specifies computer system standards. The abbreviation is often used as an adjective to computer systems that conform to these standards.

American Standard Code for Information Interchange (ASCII)
A set of codes for representing alphanumeric information (e.g., a byte with a value of 77 represents a capital M). Text files, such as those created with a computer system's text editor, are often referred to as ASCII files.[2]

AM/FM
See **automated mapping/facilities management.**

analog map
Any directly viewable map on which graphic symbols portray features and values; contrast with **digital map.**[4]

annotation
Text on a drawing or map associated with identifying or explaining graphics entities shown.[5]

ANSI
See **American National Standards Institute.**

application
A program or specially defined procedure, generally in addition to the standard set of basic software functions supplied by a GIS. Historically, an application was developed by the vendor or by a third party and purchased separately. Developed to perform a series of steps, these applications may create specialized reports, complex map products, or lead an operator through a decision process. Some of the more common applications are now becoming part of the basic software functions.

arc
See **line.**

arc-node structure
The coordinate and topological data structure used by some GISs. Arcs represent lines that can define linear features, the boundary of areas, or polygons. In arc-node structures, there is an implied direction to the line so that it may have a left and right side. In this way, the area bounded by the arc can also be described, and it is not necessary to double-store coordinates for arcs that define a boundary between two areas.

architecture
In computers, the architecture determines how the computer is seen by someone who understands its internal commands and instructions and the design of its interface hardware.[5]

area
A closed figure (polygon) bounded by one or more lines enclosing a homogeneous area and usually represented only in two dimensions. Examples are states, lakes, census tracts, aquifers, and smoke plumes.

ASCII
See **American Standard Code for Information Interchange.**

aspect
A position facing a particular direction. Usually referred to in compass directions such as degrees or as cardinal directions.

attribute
1. A numeric, text, or image data field in a relational database table that describes a spatial feature such as a point, line, node, area, or cell. [2]
2. A characteristic of a geographic feature described by numbers or characters, typically stored in tabular format and linked to the feature by an identifier. For example, attributes of a well (represented by a point) might include depth, pump type, location, and gallons per minute.

automated mapping/facilities management (AM/FM)
A GIS technology focused on the specific segment of the market concerned with specialized infrastructure and geographic facility information applications and management, such as roads, pipes, and wires.

axis
A reference line in a coordinate system.[5]

band
One layer of a multispectral image representing data values for a specific range of the electromagnetic spectrum of reflected light or heat. Also, other user-specified values derived by manipulation of original image bands. A standard color display of multispectral image displays three bands, one each for red, green, and blue. Satellite imagery such as Landsat TM and SPOT provide multispectral images of the earth, some containing seven or more bands.[2]

base map
A map showing planimetric, topographic, geological, political, and/or cadastral information that may appear in many different types of maps. The base map information is drawn with other types of changing thematic information. Base map information may be as simple as major political boundaries, major hydrographic data, and major roads. The changing thematic information may be bus routes, population distribution, or caribou migration routes.

benchmark tests
Various standard tests, easily duplicated, for assisting in measurement of product performance under typical conditions of use.[5]

binary large object (BLOB)
The data type of a column in a relational database management system (RDBMS) table that can store large images or text files as attributes.[2]

bit
The smallest unit of information that can be stored and processed in a computer. A bit has two possible values, 0 or 1, which can be interpreted as YES/NO, TRUE/FALSE, or ON/OFF.[2]

BLOB
See **binary large object.**

Boolean expression
1. A type of expression based upon, or reducible to, a true or false condition. A Boolean *operator* is a key word that specifies how to combine simple logical expressions into complex expressions. Boolean operators negate a predicate (NOT), specify a combination of predicates (AND), or specify a list of alternative predicates (OR). For example, the use of AND in "DEPTH > 100 and GPM > 500."
2. Loosely, but erroneously, used to refer to logical expressions such as "DEPTH greater than 100."[2]

breakline
A line that defines and controls the surface behavior of a triangulated irregular network (TIN) in terms of smoothness and continuity. Physical examples of breaklines are ridge lines, streams, and lake shorelines.

buffer
A zone of a given distance around a physical entity, such as a point, line, or polygon.

bundled
Refers to the way software is sold. In the early days of computers, software products were sold integrated with hardware, that is, they were "bundled." *See* **unbundled.**[5]

byte
A group of contiguous bits, usually eight, that is a memory and data storage unit. For example, file sizes are measured in bytes or megabytes (1 million bytes). Bytes contain values of 0 to 255 and are most often used to represent integer numbers or ASCII characters (e.g., a byte with an ASCII value of 77 represents a capital M). A collection of bytes (often 4 or 8 bytes) is used to represent real numbers and integers larger than 255.[2]

CAD
See **computer-aided design.**

cadastre
A record of interests in land, encompassing both the nature and extent of interests. Generally, this means maps and other descriptions of land parcels as well as the identification of who owns certain legal rights to the land (such as ownership, liens, easements, mortgages, and other legal interests). Cadastral information often includes other descriptive information about land parcels.[3]

CAE
See **computer-aided engineering.**

CAM
See **computer-aided mapping.**

cardinal
Refers to one of the four cardinal directions—north, south, east, or west.

Cartesian coordinate system
A concept from French philosopher and mathematician Rene Descartes (1596-1650). A system of two or three mutually perpendicular axes along which any point can be precisely located with reference to any other point; often referred to as x, y, and z coordinates.[5] Relative measures of distance, area, and direction are constant throughout the system.

cell
The basic element of spatial information in a grid data set. Cells are always square. A group of cells forms a grid.

centroid
The "center of gravity" or mathematically exact center of a regularly or irregularly shaped polygon; often given as an x,y coordinate of a parcel of land.[5]

chain
See **line.**

character
1. A letter, number, or special graphic symbol (*, @, -) treated as a single unit of data.
2. A data type referring to text columns in an attribute table (such as NAME).[2]

clip
The spatial extraction of physical entities from a GIS file that reside within the boundary of a polygon. The bounding polygon then works much like a cookie cutter.

cluster
A spatial grouping of geographic entities on a map. When these are clustered on a map, there is usually some phenomenon causing a relationship among them (such as incidents of disease, crime, pollution, etc.).

COGO
See **coordinate geometry.**

column
A vertical field in a relational database management system (RDBMS) data file. It may store one to many bytes of descriptive information.

command
An instruction, usually one word or concatenated words or letters, that performs an action using the software. A command may also have extra options or parameters that define more specific application of the action.

computer-aided design (CAD)
A group of computer software packages for creating graphic documents.

computer-aided engineering (CAE)
The integration of computer graphics with engineering techniques to facilitate and optimize the analysis, design, construction, nondestructive testing, operation, and maintenance of physical systems.[5]

computer-aided mapping (CAM)
The application of computer technology to automate the map compilation and drafting process. Not to be confused with the older usage, computer-aided manufacturing; usually associated with CAD, as in CAD/CAM.[5]

configuration
The physical arrangement and connections of a computer and its related peripheral devices. This can also pertain to many computers and peripherals.

conflation
A set of functions and procedures that aligns the arcs of one GIS file with those of another and then transfers the attributes of one to the other. Alignment precedes the transfer of attributes and is most commonly performed by rubber-sheeting operations.[2]

conformality
Small areas on a map are represented in their true shape and angles are preserved—a characteristic of a map projection.

connectivity
1. The ability to find a path or "trace" through a network from a source to a given point. For example, connectivity is necessary to find the path along a network of streets to find the shortest or best route from a fire station to a fire.
2. A topological construct.

contiguity
The topological identification of adjacent polygons by recording the left and right polygons of each arc.

continuous data
Usually referenced to grid or raster data representing surface data such as elevation. In this instance, the data can be any value, positive or negative. Sometimes referred to as real data. In contrast, *see* **discrete data**.

contour
A line connecting points of equal value. Often in reference to a horizontal datum such as mean sea level.

conversion
1. The translation of data from one format to another (e.g., TIGER to DXF; a map to digital files).
2. Data conversion occurs when transferring data from one system to another (e.g., SUN to IBM).
3. *See* **data automation**.

coordinate
The position of a point in space with respect to a Cartesian coordinate system (*x, y*, and/or *z* values). In GIS, a coordinate often represents locations on the earth's surface relative to other locations.

coordinate geometry (COGO)
A computerized surveying-plotting calculation methodology created at MIT in the 1950s.[5]

coordinate system
The system used to measure horizontal and vertical distances on a planimetric map. In a GIS, it is the system whose units and characteristics are defined by a map projection. A common coordinate system is used to spatially register geographic data for the same area. *See* **map projection.**

coterminous
Having the same or coincident boundaries. Two adjacent polygons are coterminous when they share the same boundary (such as a street centerline dividing two blocks).[3]

curve fitting
An automated mapping function that converts a series of short, connected straight lines into smooth curves to represent entities that do not have precise mathematical definitions (such as rivers, shoreline, and contour lines).[3]

dangling arc
An arc having the same polygon on both its left and right sides and having at least one node that does not connect to any other arc.[2]

data
A general term used to denote any or all facts, numbers, letters, and symbols that refer to or describe an object, idea, condition, situation, or other factors. May be line graphics, imagery and/or alphanumerics. It connotes basic elements of information which can be processed, stored, or produced by a computer.[5]

data automation
Mostly the same as digitizing, but can also mean using electronic scanning for data collection.

data dictionary
A coded catalog of all data types or a list of items giving data names and structures. May be on-line (referred to as an automated data dictionary), in which case the codes for the data types are carried in the database. Also referred to as DD/D for data dictionary/directory.[2]

data integration
The combination of databases or data files from different functional units of an organization or from different organizations that collect information about the same entities (such as properties, census tracts, street segments). In combining the data, added intelligence is derived.

data model
1. A generalized, user-defined view of the data related to applications.
2. A formal method for arranging data to mimic the behavior of the real-world entities they represent. Fully developed data models describe data types, integrity rules for the data types, and operations on the data types. Some data models are triangulated irregular networks (TINs), images, and georelational or relational models for tabular data.[2]

database
Usually a computerized file or series of files of information, maps, diagrams, listings, location records, abstracts, or references on a particular subject or subjects organized by data sets and governed by a scheme of organization.

Hierarchical and *relational* define two popular structural schemes in use in a GIS.[5] For example, a GIS database includes data about the spatial location and shape of geographic entities as well as their attributes.

database management system (DBMS)
1. The software for managing and manipulating the whole GIS including the graphic and tabular data.
2. Often used to describe the software for managing (e.g., input, verify, store, retrieve, query, and manipulate) the tabular information. Many GISs use a DBMS made by another software vendor, and the GIS interfaces with that software.

datum
A set of parameters and control points used to accurately define the three-dimensional shape of the earth (e.g., as a spheroid). The corresponding datum is the basis for a planar coordinate system. For example, the North American datum for 1983 (NAD83) is the datum for map projections and coordinates within the United States and throughout North America.

DBMS
See **database management system.**

DEM
See **digital elevation model.**

densify
A process of adding vertices to arcs at a given distance without altering the arc's shape. *See* **spline** for a different method for adding vertices.

digital
Usually referring to data that is in computer-readable format.

digital elevation model (DEM)
1. A raster storage method developed by the U.S. Geological Survey (USGS) for elevation data.
2. The format of the USGS elevation data sets.

digital exchange format (DXF)
1. ASCII text files defined by Autodesk, Inc. (Sausalito, Calif.) at first for CAD, now showing up in third-party GIS software.[5]
2. An intermediate file format for exchanging data from one software package to another, neither of which has a direct translation for the other but where both can read and convert DXF data files into their format. This often saves time and preserves accuracy of the data by not reautomating the original.

digital line graph (DLG)
1. In reference to data, the geographic and tabular data files obtained from the U.S. Geological Survey (USGS) that may include base categories such as transportation, hydrography, contours, and public land survey boundaries.
2. In reference to data format, the formal standards developed and published by the USGS for exchange of cartographic and associated tabular data files. Many non-DLG data may be formatted in DLG format.

digital map
A machine-readable representation of a geographic phenomenon stored for display or analysis by a digital computer; contrast with **analog map**.[4]

digital terrain model (DTM)
A computer graphics software technique for converting point elevation data into a terrain model displayed as a contour map, sometimes as a three-dimensional "hill and valley" grid view of the ground surface.[5]

digitize
A means of converting or encoding map data that are represented in analog form into digital information of x and y coordinates.

digitizer
1. A device used to capture planar coordinate data, usually as x and y coordinates from existing analog maps for digital use within a computerized program such as a GIS. Also called a *digitizing table*.
2. A person who digitizes.

DIME
See **geographic base file/dual independent map encoding.**

Dirichlet tesselation
See **Thiessen polygons.**

discrete data
Categorical data such as types of vegetation or class data such as speed zones. In geographic terms, discrete data can be represented by polygons. Sometimes referred to as integer data. In contrast, *see* **continuous data.**

distributed processing
Where computer resources are dispersed or distributed in one or more locations. The individual computers in a distributed processing environment can be linked by a communications network to each other and/or to a host or supervisory computer.

DLG
See **digital line graph.**

dots per inch (dpi)
Often referred to in printing/plotting processes and relates to how sharply an image may be represented. More dots per inch implies that edges of images are more precisely represented.

double-precision
Refers to a level of coordinate accuracy based on the possible number of significant digits that can be stored for each coordinate. Whereas single-precision coverages can store up to 7 significant digits for each coordinate and thus retain a precision of 1 meter in an extent of 1,000,000 meters, double-precision coverages can store up to 15 significant digits per coordinate (typically 13 to 14 significant digits) and therefore retain the accuracy of much less than 1 meter at a global extent.[2]

dpi
See **dots per inch.**

DTM
See **digital terrain model.**

DXF
See **digital exchange format.**

eastings
The *x*-coordinates in a plane-coordinate system; *see* **northings.**

edge match
An editing procedure to ensure that all features crossing adjacent map sheets have the same edge locations, attribute descriptions, and feature classes.

feature
A representation of a geographic entity, such as a point, line, or polygon.

file
A single set of related information in a computer that can be accessed by a unique name (e.g., a text file created with a text editor, a data file, a DLG file). Files are the logical units managed on disk by the computer's operating system. Files may be stored on tapes or disks.[2]

flat file
A structure for storing data in a computer system in which each record in the file has the same data items or fields. Usually, one field is designated as a "key" that is used by computer programs for locating a particular record or set of records or for sorting the entire file in a particular order.[3]

font
A logical set of related patterns representing text characters or point symbology (e.g., A,B,C). A font pattern is the basic building block for markers and text symbols.[2]

foreign key
In relational database management system (RDBMS) terms, the item or column of data that is used to relate one file to another.

format
1. The pattern in which data are systematically arranged for use on a computer.
2. A file format is the specific design of how information is organized in the file. For example, DLG, DEM, and TIGER are geographic data sets in particular formats available for many parts of the United States.

Fourier analysis
A method of dissociating time series or spatial data into sets of sine and cosine waves.[1]

fractal
An object having a fractional dimension; one that has variation that is self-similar at all scales, in which the final level of detail is never reached and never can be reached by increasing the scale at which observations are made.[1]

gap
The distance between two objects that should be connected. Often occurs during the digitizing process or in the edge-matching process.

GBF/DIME
See **geographic base file/dual independent map encoding.**

generalize
1. Reduce the number of points, or vertices, used to represent a line.
2. Increase the cell size and resample data in a raster format GIS.

geocode
The process of identifying a location as one or more x,y coordinates from another location description such as an address. For example, an address for a student can be matched against a TIGER street network to locate the student's home.[2]

geographic base file/dual independent map encoding (GBF/DIME)
A data exchange format developed by the U.S. Census Bureau to convey information about block-face/street address ranges related to 1980 census tracts. These files provide a schematic map of a city's streets, address ranges, and geostatistical codes relating to the Census Bureau's tabular statistical data. *See also* **topologically integrated geographic encoding and referencing (TIGER) data** created for the 1990 census.

geographic data
The composite locations and descriptions of geographic entities.

geographic database
Efficiently stored and organized spatial data and possibly related descriptive data.

geographic information retrieval and analysis (GIRAS)
Data files from the U.S. Geological Survey (USGS). GIRAS files contain information for areas in the continental United States, including attributes for land use, land cover, political units, hydrologic units, census and county subdivisions, and federal and state landownerships. These data sets are available to the public in both analog and digital form.

geographic information system (GIS)
An organized collection of computer hardware, software, geographic data, and personnel designed to efficiently capture, store, update, manipulate, analyze, and display all forms of geographically referenced information. Certain complex spatial operations are possible with a GIS that would be very difficult, time-consuming, or impractical otherwise.[2]

geographic object
A user-defined geographic phenomenon that can be modeled or represented using geographic data sets. Examples of geographic objects include streets, sewer lines, manhole covers, accidents, lot lines, and parcels.[2]

geographical resource analysis support system (GRASS)
1. A public-domain raster GIS modeling product of the U.S. Army Corp of Engineer's Construction Engineering Research Laboratory (CERL).
2. A raster data format that can be used as an exchange format between two GISs.

georeference
To establish the relationship between page coordinates on a paper map or manuscript and known real-world coordinates.[2]

GIRAS
See **geographic information retrieval and analysis.**

GIS
See **geographic information system.**

graduated circle
A circular symbol whose area, or some other dimension, represents a quantity.

graphical user interface (GUI)
A graphical method used to control how a user interacts with a computer to perform various tasks. Instead of issuing commands at a prompt, the user is presented with a "dashboard" of graphical buttons and other functions in the form of icons and objects on the display screen. The user interacts with the system using a mouse to point-and-click. For example, press an icon button and the function is performed. Other GUI tools are more dynamic and involve things like moving an object on the screen, which invokes a function. For example, a slider bar is moved back and forth to determine a value associated with a parameter of a particular operation (e.g., setting the scale of a map).[2]

GRASS
See **geographical resource analysis support system.**

graticule
A grid of parallels and meridians on a map.[4]

grid data
1. One of many data structures commonly used to represent geographic entities. A raster-based data structure composed of square cells of equal size arranged in columns and rows. The value of each cell, or group of cells, represents the entity value.
2. A set of regularly spaced reference lines on the earth's surface, a display screen, a map, or any other object.
3. A distribution system for electricity and telephones.

GUI
See **graphical user interface.**

hardware
Components of a computer system, such as the CPU, terminals, plotters, digitizers, or printers.

hierarchical
This type of data storage refers to data linked together in a treelike fashion, similar to the concept of family lines, where data relations can be traced through particular arms of the hierarchy. These data are dependent on the data structure.

hierarchy
Refers to information that has order and priority.

IGES
See **initial graphics exchange specification.**

image
A graphic representation or description of an object that is typically produced by an optical or electronic device. Common examples include remotely sensed data such as satellite data, scanned data, and photographs. An image is stored as a raster data set of binary or integer values representing the intensity of reflected light, heat, or another range of values on the electromagnetic spectrum. Remotely sensed images are digital representations of the earth.[2]

impedance
The amount of resistance (or cost) required to traverse through a portion of a network such as a line, or through one cell in a grid system. Resistance may be any number of factors defined by the user such as travel distance, time, speed of travel times the length, slope, or cost.

index
A specialized lookup table or structure within a database and used by an RDBMS or GIS to speed searches for tabular or geographic data.

infrastructure
The fabric of human improvements to natural settings that permits a community, neighborhood, town, city, metropolis, region, state, etc. to function.

initial graphics exchange specification (IGES)
An interim standard format for exchanging graphics data among computer systems.[1]

integer
A number without a decimal. Integer values can be less than, equal to, or greater than zero.

integrated terrain unit mapping (ITUM)
The process of adjusting terrain unit boundaries so there is increased coincidence between the boundaries of interdependent terrain variables such as hydrography, geology, physiography, soils, and vegetation units. Often, when this is performed, one layer or unit of geographical/descriptive information contains more than one central theme.

intelligent infrastructure
The result of automating infrastructure information management using modern computer image and graphics technology integrated with advanced database management systems (DBMSs); used for spatially linked and networked facilities and land records systems. In addition, intelligent infrastructure systems manage work processes that deal with design, construction, operation, and maintenance of infrastructure elements.[5]

item
A field or column of information within an RDBMS.

ITUM
See **integrated terrain unit mapping.**

jaggies
A jargon term for curved lines that have a stepped or saw-tooth appearance on a display device.[1]

join
To connect two or more separate geographic data sets.

key
An item or column within an RDBMS that contains a unique value for each record in the database.

kriging
An interpolation technique based on the premise that spatial variation continues with the same pattern.

LAN
See **local area network.**

lat/long
See **latitude; longitude**

latitude
A method to measure the earth representing angles of a line extending from the center of the earth to the earth's surface. With the equator representing $0°$, angles are measured in degrees north or south until $90°$ is obtained at the north and south poles. Lines of latitude are often called *parallels.*

layer
A logical set of thematic data, usually organized by subject matter.

library
A collection of repeatedly used items such as a symbol library—often-used graphics objects shown on a map or often-used program subroutines.[5]

line
1. A set of ordered coordinates that represents the shape of a geographic entity too narrow to be displayed as an area (e.g., contours, street centerlines, and streams). A line begins and ends with a node.
2. A line on a map (e.g., a neatline).

local area network
Computer data communications technology that connects computers at the same site. Computers and terminals on a LAN can freely share data and peripheral devices, such as printers and plotters. LANs are composed of cabling and special data communications hardware and software.[2]

longitude
A method to measure the earth representing angles of a line extending from the center of the earth to the earth's surface. With a line extending from the north to the south poles and passing through Greenwich, England, as $0°$, angles are measured in degrees east or west until $180°$s is obtained at the opposite side of the earth from $0°$ longitude. Lines of longitude are often called meridians.

macro
A set of instructions used by a computer program or programs. These are usually stored in a text file and invoked from a program that reads this text file as if the commands were typed interactively.

many-to-one relate
A relate in which many records can be related to a single record. A typical goal in relational database design is to use many-to-one relates to reduce data storage and redundancy.[2]

map projection
A mathematical model for converting locations on the earth's surface from spherical to planar coordinates, allowing flat maps to depict three-dimensional features. Some map projections preserve the integrity of shape; others preserve accuracy of area, distance, or direction.[2]

map units
The coordinate units in which the geographic data are stored, such as inches, feet, meters, or degrees, minutes and seconds.

meridian
A line running vertically from the north pole to the south pole along which all locations have the same longitude. The prime meridian (0°) runs through Greenwich, England. Moving left or right of the prime meridian, measures of longitude are negative to the west and positive to the east up to 180° halfway around the globe.[2]

metropolitan statistical area (MSA)
A single county or group of contiguous counties that define a metropolitan region, usually with a central city with at least 50,000 inhabitants; in the past these have been called standard metropolitan statistical areas (SMSAs) and standard metropolitan areas (SMAs); the precise definitions and changes therein are established by the U.S. Office of Management and Budget.[4]

minimum bounding rectangle
The rectangle defined by the map extent of a geographic data set and specified by two coordinates: xmin, ymin and xmax, ymax.[2]

minor civil division (MSD)
The primary political or administrative subdivision of a county.[4]

model
1. An abstraction of reality. Models can include a combination of logical expressions, mathematical equations, and criteria that are applied for the purpose of simulating a process, predicting an outcome, or characterizing a phenomenon. The terms *modeling* and *analysis* are often used interchangeably, although the former is more limited in scope.
2. Data representation of reality (e.g., spatial data models include the arc-node, georelational model, rasters or grids, and TINs).

neatline
A border line commonly drawn around the extent of a map to enclose the map, legend, scale, title, and other information, keeping all of the information pertaining to that map in one "neat" box.

network
1. A system of interconnected elements through which resources can be passed or transmitted, for example, a street network with cars as the resource or electric network with power as the resource.
2. In computer operations, the means by which computers connect and communicate with each other or with peripherals.

network analysis
The technique utilized in calculating and determining relationships and locations arranged in networks, such as in transportation, water, and electrical distribution facilities.[5]

node
1. The beginning or ending location of a line.
2. The location where lines connect.
3. In graph theory, the location at which three or more lines connect.
4. In computers, the point at which one computer attaches to a communication network.

northings
The y-coordinates in a plane-coordinate system; *see* **eastings**.[4]

operating system (OS)
Computer software designed to allow communication between the computer and the user. For larger computers, it is usually supplied by the manufacturer. The operating system controls the flow of data, the interpretation of other programs, the organization and management of files, and the display of information. Commonly known OSs are VMS, VM/IS, UNIX, DOS, and OS/2.

OS
See **operating system.**

output
The results of processing data.

overshoot
That portion of a line digitized past its intersection with another line. Sometimes referred to as a dangling line.

pan
To move the spatial view of data to a different extent without changing the scale.

parallel
1. A property of two or more lines that are separated at all points by the same distance.
2. A horizontal line encircling the Earth at a constant latitude. The equator is a parallel whose latitude is 0°. For example, measures of latitude are positive up to 90° above the equator and negative below.[2]

pathname
The direction(s) to a file or directory location on a disk. Pathnames are always specific to the computer operating system. Computer operating systems use directories and files to organize data. Directories are organized in a tree structure; each branch on the tree represents a subdirectory or file. Pathnames indicate locations in this hierarchy.[2]

peripheral
A component such as a digitizer, plotter, or printer that is not part of the central computer but is attached through communication cables.

pixel
One picture element of a uniform raster or grid file. Often used synonymously with **cell**.

plane-coordinate system
A system for determining location in which two groups of straight lines intersect at right angles and have as a point of origin a selected perpendicular intersection.[4]

planimetric map
A large-scale map with all features projected perpendicularly onto a horizontal datum plane so that horizontal distances can be measured on the map with accuracy.[4]

PLSS
See **public land survey system.**

point
1. A single *x,y* coordinate that represents a geographic feature too small to be displayed as a line or area, for example, the location of a mountain peak or a building location on a small-scale map.[2]
2. Some GIS systems also use a point to identify the interior of a polygon.

polygon
A vector representation of an enclosed region, described by a sequential list of vertices or mathematical functions.

precision
1. If applied to paper maps or map databases, it means exactness and accuracy of definition and correctness of arrangement.[5]
2. If applied to data collection devices such as digitizers, it is the exactness of the determined value (i.e., the number 134.98988 is more precise than the number 134.9).
3. The number of significant digits used to store numbers.

primary key
The central item or column within an RDBMS that contains a unique value for each record in the database, such as the unique number assigned to each parcel within a county.

projection
See **map projection.**

public land survey system (PLSS)
A rectangular survey system that utilizes 6-mile-square townships as its basic survey unit. The location of townships is controlled by baselines and meridians running parallel to latitude and longitude lines. Townships are defined by range lines running parallel (north-south) to meridians and township lines running parallel (east-west) to baselines. The PLSS was established in the United States by the Land Ordinance of 1785.[3]

quadrangle
A four-sided region, usually bounded by a pair of meridians and a pair of parallels.[4]

quadtree
A spatial index that breaks a spatial data set into homogeneous cells of regularly decreasing size. Each decrement in size is one-forth the area of the previous cell. The quadtree segmentation process continues until the entire map is partitioned. Quadtrees are often used for storing raster data.

raster data
Machine-readable data that represent values usually stored for maps or images and organized sequentially by rows and columns. Each "cell" must be rectangular, but not necessarily square, as with grid data.

RDBMS
See **relational database management system.**

record
In an attribute table, a single "row" of thematic descriptors.

rectify
The process by which an image or grid is converted from image coordinates to real-world coordinates. Rectification typically involves rotation and scaling of grid cells and thus requires resampling of values.[2]

relate
An operation establishing a connection between corresponding records in two tables using an item common to both. Each record in one table is connected to one or more records in the other table that share the same value for a common item.

relational
A type of data storage involving tabular data where the storage structure is independent of the relations.

relational database management system (RDBMS)
A database management system with the ability to access data organized in tabular files that may be related together by a common field (item). An RDBMS has the capability to recombine the data items from different files, thus providing powerful tools for data usage.[2]

relational join
The process of connecting two tables of descriptive data by relating them by a key item, then merging the corresponding data. The common key item is not duplicated in this process.

resolution
1. The accuracy at which the location and shape of map features can be depicted for a given map scale. For example, at a map scale of 1:63,360 (1 inch = 1 mile), it is difficult to represent areas smaller than 1/10-mile wide or 1/10-mile in length because they are only 1/10-inch wide or long on the map. In a larger scale map, there is less reduction, so feature resolution more closely matches real-world features. As map scale decreases, resolution also diminishes because feature boundaries must be smoothed, simplified, or not shown at all.

2. The size of the smallest feature that can be represented in a surface.
3. The number of points in x and y in a grid (e.g., the resolution of a USGS one-degree DEM is 1,201 x 1,201 mesh points).[2]

route
A process that establishes connections through a network or grid from a source to a destination. A network example would be to establish a route through a network of streets from a fire station to the fire. A grid example would be to move soil particles from a ridgetop to a stream based on equations developed by soil scientists. The determination of these routes usually takes into consideration impedances.

row
1. A record in an attribute table.
2. A horizontal group of cells in a grid or pixels in an image.

rubber-sheeting
A procedure to adjust the entities of a geographic data set in a nonuniform manner. From- and to-coordinates are used to define the adjustment.

scale
The relationship existing between a distance on a map and the corresponding distance on the earth. Often used in the following form 1:24,000, which means that 1 unit of measurement on the map equals 24,000 of the same units on the earth's surface.

scanning
Also referred to as *automated digitizing* or *scan digitizing*. A process by which information originally in hard copy format (paper print, mylar™ transparencies, microfilm aperture cards) can be rapidly converted to digital raster form (pixels) using optical readers.[5]

single-precision
A lower level of coordinate accuracy based on the possible number of significant digits that can be stored for each coordinate. Single-precision numbers can store up to seven significant digits for each coordinate and thus retain a precision of ±5 meters in an extent of 1,000,000 meters. Double-precision numbers can store up to 15 significant digits (typically 13 to 14 significant digits) and therefore retain the accuracy of much less than 1 meter at a global extent.[2]

sliver polygon
A relatively narrow feature commonly occurring along the borders of polygons following the overlay of two or more geographic data sets. Also occurs along map borders when two maps are joined as a result of inaccuracies of the coordinates in either or both maps.

smoothing
A process to generalize data and remove smaller variation.

software
A computer program that provides the instructions necessary for the hardware to operate correctly and to perform the desired functions. Some kinds of software are operating system, utility, and applications.[5]

soundex
A phonetic spelling (up to six characters) of a street name, used for address matching. Each of the 26 letters in the English alphabet are replaced with a letter in the soundex equivalent:
English: A B C D E F G H I J K L M N O P Q R S T U V W X Y Z
Soundex: A B C D A B C H A C C L M M A B C R C D A B W C A C
Where possible, geocoding uses a soundex equivalent of street names for faster processing. During geocoding, initial candidate street names are found using soundex, then real names are compared and verified.[2]

spatial index
A means of accelerating the drawing, spatial selection, and entity identification by generating geographic-based indexes. Usually based on an internal sequential numbering system.

spatial model
Analytical procedures applied with a GIS. There are three categories of spatial modeling functions that can be applied to geographic data objects within a GIS: (1) geometric models (such as calculation of Euclidian distance between objects, buffer generation, and area and perimeter calculation); (2) coincidence models (such as polygon overlay); and (3) adjacency models (pathfinding, redistricting, and allocation). All three model categories support operations on geographic data objects such as points, lines, polygons, TINs, and grids. Functions are organized in a sequence of steps to derive the desired information for analysis.[2]

spike
1. An overshoot line created erroneously by a scanner and its raster software.
2. An anomalous data point that protrudes above or below an interpolated surface representing the distribution of the value of an attribute over an area.[2]

spline
A method to mathematically smooth spatial variation by adding vertices along a line. *See* **densify** for a slightly different method for adding vertices.

SQL
See **Structured Query Language.**

string
See **line.**

Structured Query Language (SQL)
A syntax for defining and manipulating data from a relational database. Developed by IBM in the 1970s, it has since become an industry standard for query languages in most RDBMSs.[2]

surface
A representation of geographic information as a set of continuous data in which the map features are not spatially discrete; that is, there is an infinite set of values between any two locations. There are no clear or well-defined breaks between possible values of the geographic feature. Surfaces can be represented by models built from regularly or irregularly spaced sample points on the surface.[2]

surface model
Digital abstraction or approximation of a surface. Because a surface contains an infinite number of points, some subset of points must be used to represent the surface. Each model contains a formalized data structure, rules, and x,y,z point measurements that can be used to represent a surface.[2]

syntax
A set of rules governing the way statements can be used in a computer language.[1]

table
1. Usually referred to as a *relational table*. The data file in which the relational data reside.
2. A file that contains ASCII or other data.

template
1. A geographic data set containing boundaries, such as land-water boundaries, for use as a starting place in automating other geographic data sets. Templates save time and increase the precision of spatial overlays.
2. A map containing neatlines, north arrow, logos, and similar map elements for a common map series but lacking the central information that makes one map unique from another.
3. An empty tabular data file containing only item definitions.

thematic map
A map that illustrates one subject or topic either quantitatively or qualitatively.[4]

theme
A collection of logically organized geographic objects defined by the user. Examples include streets, wells, soils, and streams.

Thiessen polygons
Polygons whose boundaries define the area that is closest to each point relative to all other points. Thiessen polygons are generated from a set of points. They are mathematically defined by the perpendicular bisectors of the lines between all points. A triangulated irregular network (TIN) structure is used to create Thiessen polygons.[2]

TIGER
See **topologically integrated geographic encoding and referencing data.**

tile
A part of the database in a GIS representing a discrete part of the earth's surface. By splitting a study area into tiles, considerable savings in access times and improvements in system performance can be achieved.[1]

TIN
See **triangulated irregular network.**

topographic map
A map of land-source features, including drainage lines, roads, landmarks, and usually relief or elevation.[4]

topologically integrated geographic encoding and referencing (TIGER) data
A format used by the U.S. Census Bureau for the 1990 census to support census programs and surveys. TIGER files contain street address ranges along lines and census tract/block boundaries. These descriptive data can be used to associate address information and census/demographic data to coverage features.[2]

topology
The spatial relationships between connecting or adjacent coverage features (e.g., arcs, nodes, polygons, and points). For example, the topology of an arc includes its from- and to-nodes, and its left and right polygons. Topological relationships are built from simple elements into complex elements: points (simplest elements), arcs (sets of connected points), areas (sets of connected arcs), and routes (sets of sections that are arcs or portions of arcs). Redundant data (coordinates) are eliminated because an arc may represent a linear feature, part of the boundary of an area feature or both. Topology is useful in GIS because many spatial modeling operations do not require coordinates, only topological information. For example, to find an optimal path between two points requires a list of which arcs connect to each other and the cost of traversing along each arc in each direction. Coordinates are only necessary to draw the path after it is calculated.[2]

transformation
The process of converting data from one coordinate system to another through translation, rotation, and scaling.

triangulated irregular network (TIN)
A representation of a surface derived from irregularly spaced sample points and breakline features. The TIN data set includes topological relationships between points and their proximal triangles. Each sample point has an x,y coordinate and a surface or z value. These points are connected by edges to form a set of nonoverlapping triangles that can be used to represent the surface. TINs are also called *irregular triangular mesh* or *irregular triangular surface model*.[2]

tuple
Synonym of **record**.

unbundled
Refers to software sold separately from hardware. *See* **bundled**.

undershoot
A digitized line that does not quite reach a line that it should intersect. As with an overshoot, this is also sometimes referred to as a *dangling line*.

Universal Transverse Mercator (UTM)
A widely used planar coordinate system, extending from 84° north to 80° south **latitude** and based on a specialized application of the Transverse Mercator projection. The extent of the coordinate system is broken into 60, 6-degree (longitude) zones. Within each zone, coordinates are usually expressed as meters north or south of the equator and east from a reference axis. For locations in the Northern Hemisphere, the origin is assigned a false **easting** of 500,000 and a false **northing** of 0. For locations in the Southern Hemisphere, the origin is assigned a false easting of 500,000 and a false northing of 10,000,000.

UTM
See **Universal Transverse Mercator.**

vector data
A coordinate-based data structure commonly used to represent map features. Each linear feature is represented as a list of ordered *x,y* coordinates. Attributes are associated with the feature (as opposed to a raster data structure, which associates attributes with a grid cell). Traditional vector data structures include double-digitized polygons and arc-node models.[2]

vertex
One point along a line.

z-value
The elevation value of a surface at a particular *x,y* location. Also, often referred to as a *spot value* or *spot elevation.*[2]

zoom
To display a smaller of larger region instead of the present spatial data set extent to show greater or lesser detail.

References

[1]Burrough, P.A. 1990. *Principles of Geographical Information Systems for Land Resources Assessment* (Somerset, UK: Butler & Tanner).

[2]Environmental Systems Research Institute, Inc. 1991. *ARC/INFO Data Model, Concepts, and Key Terms* (Redlands, CA: ESRI, Inc.).

[3]Huxhold, W. 1991. *An Introduction to Urban Geographic Information Systems* (New York: Oxford University Press).

[4]Monmonier, M., and G.A. Schnell. 1988. *Map Appreciation* (Englewood Cliffs, NJ: Prentice Hall).

[5]Montgomery, G., and G. Juhl. 1990. *Intelligent Infrastructure Workbook* (Fountain Hills, AZ: A-E-C Automation Newsletter).

Index

❖   ❖   ❖